BEEKEEPING

A PRACTICAL GUIDE

Richard E. Bonney

STOREY BOOKS

The mission of Storey Publishing is to serve our customers
by publishing practical information that encourages personal independence
in harmony with the environment.

Cover and text design by Michelle Arabia
Cover photographs copyright © Grant Heilman Photography
Text production by Leslie Carlson
Edited by Sandra Webb and Deborah Balmuth
Line drawings by Brigita Fuhrmann
Indexed by Northwind Editorial Services

The information in this book is true and complete to the best of our knowledge. All recommendations are made without guarantee on the part of the author or Storey Publishing. The author and publisher disclaim any liability in connection with the use of this information. For additional information, please contact Storey Books, 210 MASS MoCA Way, North Adams, MA 01247.

Storey books are available for special premium and promotional uses and for customized editions. For further information, please call Storey's Custom Publishing Department at 1-800-793-9396.

Printed in the United States by Capital City Press
15   14   13

Library of Congress Cataloging-in-Publication Data

Bonney, Richard E.
       Beekeeping : a practical guide / Richard E. Bonney.
               p.   cm.

       Includes bibliographical references (p.   ) and
index.
       ISBN 0-88266-861-7 (pb)
       1. Bee culture. I. Title.
   SF523.B66   1993
   638'.1 — dc20   92-56145
                            CIP

TABLE OF CONTENTS

INTRODUCTION

The art, the craft, the science of beekeeping cannot be covered in a single book. The volume of information is simply too great. Furthermore, it is not necessary to learn everything the first year. However, you do need a certain broad exposure and a feeling for where you are headed. This book intends to give you that exposure and that sense of direction.

Without question, this is a book for beginners, although novices who have kept bees for a season or so can also benefit from it. It will take you through the first year, from spring to spring. Use the information here as a point of departure — and keep on reading about bees.

As with any endeavor, a certain amount of misinformation about bees and beekeeping is extant. Be selective in your reading and be hesitant in taking information on faith. Know your source, ask questions, and be sure the answers make sense.

This book is based on the experience of beekeeping in the northeastern United States. In this region, the active season for a honey bee colony begins in late April and extends through the end of September. Bees in other parts of the country have the same basic cycle, with adjustments in the calendar to reflect the climate and the specific forage available.

CHAPTER 1

BEFORE WE START

Most new beekeepers come into this exciting endeavor as hobbyists. Their whole attitude toward beekeeping is colored by this approach — it's relaxed, casual, and looks like fun — let's try it out. Beekeeping can be an enjoyable hobby. However, it can also be a disappointing failure. Beekeeping requires preparation, ongoing attention, and commitment. It requires knowledge — of bees, of growing things, and about the natural world in general. Beekeeping requires a certain amount of interaction with others in the beekeeping world, although this does not have to be an involved interaction.

The Beekeeper's Commitment

Too many novice beekeepers do not recognize the level of commitment they must have. They do not know that, over the long run, for every beekeeper who succeeds, there are probably two or three who do not. How many of you have been in a classroom or a training program where the instructor says at the outset, "Look at the person on your right; now look at the person on your left. One

of you won't be here next year (or next week or next month)."
Beekeeping is like that.

Furthermore, not all beekeepers are truly bee**keepers.** Some
are bee**havers**. They develop an initial enthusiasm, acquire some
bees, perhaps learn about them and work with them for a while,
and then lose interest. Or perhaps they never develop any real
knowledge or enthusiasm at all. One way or the other they become
beehavers: they have some bees in the backyard but are not truly
keeping them. The bees keep themselves. All may be well for a
while, but in the long run, this seldom has a happy ending.

To be a successful beekeeper means to be a committed beekeeper, one who learns about the bees, comes to understand them,
works with them on a regular basis, and enjoys them. If you believe
your involvement might be something less than this, perhaps beekeeping is not for you.

It is difficult to keep bees if you don't have contact at some level
with other beekeepers and with sources of new developments —
government agencies; universities; and local, state, or national beekeeper organizations. All of these sources are an important part of
the overall beekeeping picture. Contact can be as simple as mem-
bership in a beekeepers' club or association or regular reading of a
beekeepers' magazine. Taking both approaches is not too much.
Beekeeping is a dynamic endeavor. Problems arise, solutions are
worked out, research is undertaken, new knowledge comes to the
fore continually. Knowledge of bees and of beekeeping has increased immensely in recent years, as have the problems. A beekeeper who is out of touch will be quickly overwhelmed.

Some individuals who take up beekeeping do so because they
need bees for pollination. They want bees on their property to take
care of their crops, whether it be a small home garden or a commercial operation. They need the bees to ensure an adequate crop.
Their approach is to obtain one or several colonies, put the bees out
in a corner of the property, and forget about them, assuming they
will take care of themselves. And the bees will, for a while. But the
facts of bee life — disease, drought, an unusually harsh winter,
predators, any or all of these problems and others — can cause the
colony to weaken and to die. It happens regularly in nature. We
don't think about it because we don't see it, but a colony of honey

bees often has a tenuous grip on life in North America, especially in the more northerly regions. Feral colonies die regularly. They are replaced almost as regularly by swarms from other feral colonies or from a beekeeper's holdings. But this is not a desirable situation. We are not in control, and with the problems mentioned earlier, we must be. Someone who wants bees on his or her property, but does not wish to care for them, should think seriously of finding a bee-keeper who is willing to establish colonies on the property in question: It is a much more satisfactory approach in the long run. With all of these thoughts in mind, let's now prepare to get started.

Dimensions of Beekeeping

Beekeeping is multifaceted. It is much more than placing a hive in the backyard, visiting it a couple of times a year, and reaping the benefits in terms of honey and pollination. I will discuss the actual benefits, the returns, later on, but for now, what must go into this new endeavor? What are the dimensions, the scope, of the activity you are about to undertake? How much time is involved? What is the cost? In what ways will it be restrictive? We will address each of these in turn.

Time

The time devoted to keeping bees does not have to be great. It may be only a few hours per year — once you are past the initial learning period. Of course, it also might be many hours. As with most endeavors, what you get back is proportionate to what you put in. Furthermore, some people have different goals when they take up with bees.

No matter what your goals, the first time commitment should be to learning. This is accomplished by reading, attending an informal bee school and a workshop or two, attending occasional club or association meetings, and talking with other beekeepers. Many in-dividuals undertake beekeeping with a minimum of preparation, believing that they can just dive in and pick up the requisite knowl-edge on the fly. It doesn't work that way. Success comes only after the acquisition of basic knowledge. Plan on a commitment of learning time that will be heavy during the first year or so and less as time passes.

Given a basic level of understanding and a willingness to continue learning at a modest level, you should plan on visiting your bees at least once every 2 weeks during the active season, perhaps more often in the spring as the new season is getting underway and certainly less often in the winter. However, bees should never be totally ignored, even in winter. Life goes on in the colony year-round.

Individual visits per hive may be quite brief, depending on the season and your reason for being at the hive. Some visits may last a minute or two; others involving a specific task may last 20 to 30 minutes per hive, although usually not longer. Most visits that involve opening the hive are a substantial disruption to colony life but serve an important purpose. Although you can skip visits now and then (beekeepers are allowed to go on vacation), ignoring the bees for weeks on end only leads to problems. Some endeavors or hobbies can be picked up or put down at will. Beekeeping is different. Chores not done at the proper time usually cannot be done as well later, if they can be done at all.

All in all, the time involved in keeping bees is not great, once you are past the first year. The dimensions of the involvement should become clearer as you progress through the book. But the intensity of this effort deserves some thought.

Money

Why are you going to keep bees? Most new beekeepers see it as an interesting hobby, and if they can make money on the side, so much the better. Others go into it with the specific intent of making money: a new beekeeper seldom does. An experienced beekeeper may, but no one should count on it. Far too many variables and problems exist for any beekeeper to ever be completely in control. Go into beekeeping strictly as a hobby. Expect it to cost money. Initially, it will be all outgo. If, after a couple of years, you have surplus honey to sell — that's wonderful. In time, you may even have enough hives to go into crop pollination in a small way. Eventually, if all is going well, perhaps you can turn this hobby into a sideline business, but don't base your future on it now. Wait until you see what it's all about.

So, how is the first season going to treat your pocketbook? I'll throw out a ballpark figure now and address this topic in a little more detail in the section on equipment. For now, assume an expenditure of $150 to $200, depending on sources and quality, to set up a complete hive with a couple of honey supers, bees included. Necessary additional equipment, such as smoker, hive tool, veil, and so on, may cost another $60 to $75. Although in the past it was sometimes possible to recover part of this investment through honey production during the first or second season, it is highly unlikely that you can do so in today's economy. In recent years, the cost of beekeeping equipment and supplies has risen much, much faster than the value of honey.

The Down Side

Beekeeping is not all pleasure. It does have a down side — as most things in life do.

Some beekeepers have a partner and they work their hives together. However, most beekeepers are loners. You will be out there by yourself, in the heat, sticky to your elbows, bees buzzing about, a veil in place so you can't scratch or blow your nose or take a drink of water. Occasionally, you will infuriate the bees, and they will find ways to get under your veil, up your sleeves, or into some other place you might not want to discuss in polite company. There is nothing more disconcerting than to realize that a bee is inside your pants, crawling up your leg, and already at knee level. As you move about, your clothing is eventually going to pinch the bee and cause it to sting. You have two choices: leave the beeyard and disrobe carefully so the bee can escape, or slap where the bee is and kill it. It won't go away by itself.

Do not discount the heat, the weight of the equipment, and the discomfort of work in a beeyard. These discomforts are not a regular feature of beekeeping, but there are times when you will find yourself wondering what you are doing in this position, the hive open and its parts strewn about you, bees everywhere, sweat streaming — and suddenly, you realize that there is a bee inside the veil. But like so many negative experiences, it will be soon forgotten. You will go back there again and have a great time. Just be aware that life in the beeyard does have its challenges.

The Up Side

Without question, there is an up side to beekeeping that far outweighs any down side. Why else have thousands of individuals persisted with it for so many years? Go to any gathering of beekeepers and listen to them talk with enthusiasm, even about their problems. Go to a beeyard on a pleasant summer day and sit there. Immerse yourself in the calmness and serenity of the scene, watch the bees coming and going, sometimes indulging in so-called play time. If beekeeping is truly for you, you will not be able to resist.

Stings

Honey bees sting. Every one knows this. It is a fact of life. In the beekeeper's scheme of things, stings are usually but a minor irritation. Consider that if bees could not sting, we would probably not have any particular interest in them. Stinging is a defensive behavior. It is this defensive behavior that allows a colony of bees to store away large quantities of honey with minimum likelihood that it will be taken from them by predators. They are able to defend themselves, their home, and their wealth. Because they can store surplus honey, bees can survive the winter in relatively large numbers and can greet the new growing season ready to work. Under normal circumstances, that work results in more stores of honey — some for us, some for them. So stinging has a positive side. But it can create problems.

A bee sting itself is an interesting event, not a particularly scary one. A single sting, although momentarily painful, normally has no persistent aftereffects. A beekeeper probably does not even experience any distinct pain. I will discuss this later, but first, let's consider bee sting reactions, especially as perceived by the general public.

BEE STINGS

Some people find it hard to believe that stinging is a defensive action when apparently they have been doing nothing to stimulate a stinging incident. A closer analysis of the event, however, usually uncovers the stimulus. A bee stings in defense of its person or its home. A bee is not a thinking creature but a reactive one. If something takes place that the bee perceives to be an attack, it reacts. Sticking your nose into a flower where a bee is sipping nectar is perceived by the bee as an aggressive action, even though your intentions are benign. Swatting at a bee that flies near you is clearly an aggressive act and can trigger a defensive reaction in the bee. Walking too close to a hive can also be perceived as aggression by the bees, especially if the hive has been subjected to some kind of harassment in the recent past—vandalism, for instance, or animal depredation. Some animals are willing to put up with a few stings and persistently annoy the bees to get at the goodies in the hive. In such instances, the bees are especially reactive for hours or even days later. And finally, there are people who are stung for no other reason than they simply are there. Either their body chemistry annoys the bees or they are wearing a fragrance (perfume, deodorant, or hair dressing, for example) that is attractive or offensive to the bees. These people walk near a colony of bees and find themselves being stung. Bees' lives are governed by odors and their "noses" are very sensitive.

Many people believe they are allergic to bee stings. Some have been told by their physicians that they are allergic and to consider any insect sting life threatening. Often, these people are not truly allergic. They may suffer from swelling and pain but these are confined to the single area of the sting site. It is better to be safe than sorry. The line between local and systemic reactions is often quite fuzzy and physicians deliberately (and properly) err on the side of caution. Furthermore, although each successive sting usually elicits a milder reaction as the body becomes resistant, it can work the other way, with subsequent stings being more serious.

Reaction to Stings

We can consider three general levels of bee sting reaction. In the first, the victim is stung, the bee departs, the stinger is removed, and there are few, if any, aftereffects. This is the usual beekeeper reaction. There may be pain, intense or minimal, for a minute or so, and a welt left by the sting. Perhaps the sting site itches. Usually, the incident is over and forgotten soon, although some itching or minor discomfort may persist for several hours. Most people react to a bee sting in this way.

In the second level of reaction, swelling may take place. The swelling may be severe, startling, even scary. It may persist for a day or two. Although disturbing, unpleasant, and perhaps painful, this reaction alone should not be viewed as life threatening. It is generally considered to be a local reaction and it will go away. I do not believe it is properly termed an allergic reaction. (In my own case, I react this way to a single sting once every year or so. Otherwise, my reactions are minor, forgotten in a moment or two.)

The third level of reaction is of considerably more concern. The initial reaction to the sting may be as in either of the two levels just described — a welt or a serious swelling. Then other symptoms follow: nausea, body rash, difficulty in breathing, and dizziness all suggest the onset of anaphylactic shock, a life-threatening situation. Medical help must be sought immediately. People who suffer this level of reaction *should* consider themselves allergic and should either stay away from bees completely or at the very least carry a sting kit prescribed by their physician. It is also possible to have a series of shots that desensitize the body to stings.

Tolerance to Stings

Beekeepers who are stung regularly usually build up a tolerance to stings. Two things happen: First, the beekeeper, after being stung a few times, realizes that stings are not such a big deal and tends to ignore them; second, the body builds a tolerance to stings and no longer reacts as strongly. The beekeeper flicks the sting out and continues work. To build and maintain such a tolerance in the body requires that the individual be stung periodically — about once every 10 days seems to be effective. After a winter with no stings, the tolerance may wear off and have to be reestablished. A few stings in the early season will take care of this.

For some people, there is a different outcome for stings. Instead of building a tolerance, they begin to react more and more to each successive stinging incident, ultimately developing a true allergy and requiring medical assistance. This can happen to beekeepers and nonbeekeepers alike, and has been observed in members of beekeepers' families, even those who do not normally work with the bees. Apparently, the repeated breathing of particles of dried venom or other bee-related materials, which adhere to the beekeeper's clothing and are carried into the house, contributes to this type of increasing intolerance. One adverse reaction may not be enough to positively diagnose an allergic condition, but again, it is better to seek medical advice and to be safe rather than sorry.

Effect of Sting on the Bee

A sting, of course, has an effect on the bee — it is fatal. A bee's stinger is barbed and has moving parts. When the bee stings, the stinger penetrates the victim's skin and is held there by the barbs. The bee pulls away and leaves the stinger in the victim. Unfortunately for the bee, a portion of its anatomy is left attached to the stinger — specifically, the venom sac and the related glands and muscles. The bee is able to fly away but will die, perhaps in minutes, perhaps in hours. The bee may continue to fly about the victim, buzzing and threatening to sting again. Of course, it cannot sting again — it no longer has a stinger — but the victim does not necessarily know which

After a sting, the bee pulls away, leaving behind the stinger and part of its anatomy (the venom sac and related glands and muscles). It flies away, but soon dies.

bee has stung and may react by swatting, perhaps provoking other bees to sting. In addition to the stinger left in place, the bee also has left behind an alarm odor, which is a stimulus for other bees to sting.

Coping with Stings

Meanwhile, the stinger is doing its work. The stinger itself has two moving parts, or lancets, which make up the shaft. When the bee stings, the shaft penetrates and is held by the barbs near the tip. The pulsating action of the victim's muscles cause one lancet to then penetrate deeper. This first lancet is held in place by the barbs, acting as an anchor while the second one penetrates. The muscles continue to work the shaft progressively deeper. All the while, venom is being pumped from the attached sac into the wound through the center of the shaft. You can short-circuit this action by removing the stinger when you first feel the sting. By doing this quickly, you will receive a minimum dose of venom. Remove the stinger by scraping it out with a fingernail or your hive tool, for instance. Do not grab the stinger directly and pull. Doing so will squeeze the venom sac and further inject you with venom. Finally, a puff of smoke on the sting site will help to mask the alarm odor.

Of course, if the sting is persistently painful, you may want to treat the symptoms. Remedies abound. You can buy lotions made for the purpose, but usually any of the folk remedies you hear about will help. For instance, a dab of mud, a paste of baking soda and water, an ice cube, a drop or two of honey — these and many more will give relief until the pain stops.

One last thought: stings do hurt children more. It may be at least partly psychological, although their smaller size probably is a factor as well. No matter what the reason, most children genuinely seem to hurt more than older folks. Put something on the sting site quickly and give lots of sympathy.

Legal Considerations

Most states have an apiary inspection program that regulates bee-keeping to some extent. The scope of this regulation is slowly expanding in many places, and some municipalities have now imposed their own codes, primarily in the form of bans or restrictions. Historically, beehives have been regulated for the detection and the control of honey bee brood diseases. One disease, in particular, is highly contagious among bee colonies and can cause their demise. Today, we are also concerned with controlling the spread of two species of parasitic mites and of the Africanized bee, all of which have entered this country in recent years, are spreading, and are creating severe problems. Disease and mites can and do kill honey bee colonies: left unchecked, they can be devastating. The Africanized bee is a potentially serious threat to humans.

Although such problems are increasing, the condition of the economy in recent years has caused both state and federal budgets to be cut in many areas. In some instances, these cuts have had a severe impact on bee inspection programs. Some states have had their inspection programs reduced, whereas others have had them cut entirely. More pressure has been put on individual beekeepers first to recognize, then to control or to eliminate, disease and mites. This is perhaps as it should be, but many beekeepers do not realize, or do not accept, that their world is changing. They still expect some higher authority to solve all of their problems. We will be able to continue keeping bees successfully only after all beekeepers come to understand the state of beekeeping today, and accept their responsibilities.

Regulations about Bees

No federal laws exist in this country that regulate beekeeping, except the one that bans the importation of live bees. The parasitic mites mentioned above are not native to the Western Hemisphere and have been in North America only since the mid-1980s. We do not know specifically how they came into this country, but they were probably brought in illegally with queen bees — a violation of this federal ban.

Laws or regulations governing bees do exist at the state level and vary from state to state. Some states totally ban importation of live bees. Others allow importation, but only when specific conditions are met (for instance, a certificate of inspection from the originating state showing that the bees were inspected within a certain recent period and are free of disease and mites). As the Africanized bee spreads, further regulation is probable. Some moves have been made toward more uniform regulations from state to state, but little progress has been made.

It is the responsibility of each individual beekeeper to be aware of the specific laws and regulations of the state(s) in which he or she keeps bees and to follow these laws and regulations faithfully. Too much is at stake to allow anyone to go his or her own way. Mites, undetected or untreated, can wipe out a beekeeper's holdings. Seemingly unprovoked stinging attacks, as from Africanized bees or from any bees that the public may perceive to be Africanized, can result in a total ban of bees from specified areas. Unfortunately, the public and the representatives of many official agencies who should be better informed, do not understand the scope and importance of honeybees to our way of life. Perhaps we will someday have alternatives, but for now we cannot do without honey bees. It is not an exaggeration to say that our present diet, a large part of our environment, and, in turn, our way of life is dependent on the pollination services of honeybees.

Your mentor, if you have one, your local beekeeping organization, or your local equipment supplier should be able to help you find out about the laws and regulations in your state and municipality. Other sources include your state's department of agriculture and the cooperative extension system.

History of Beekeeping

The history of beekeeping may sound like a rather dull topic to someone who is anxious to get some bees and begin. However, certain of our practices, techniques, and current attitudes result from or are carried over from the earlier days of beekeeping. It is thus helpful to know at least a smattering of the history of the craft.

Bees have been kept, or at least exploited, as far back as we have records. Primitive people robbed bees of their honey, as is evidenced by ancient cave paintings. Stone-Age people probably kept bees in some manner and can be considered the first beekeepers. After that early period, we can identify at least three eras of beekeeping: the first, from ancient times until about 1500; the second, from about 1500 until 1850; and the third, from 1850 until the present. Perhaps we are now actually in a fourth era, one marked by extensive research and resultant discoveries, by the ever-increasing role of bees as pollinators of commercial crops, by the pressures of human growth and population, and by the problems of the day — mites and Africanized bees.

One of the earliest hives was made from a section of naturally hollow log. Such hives could still be found in use this century.

The First Era

During the first era, pre-1500, beekeeping was more properly beehaving; that is, people had, rather than kept, bees. The term still has application today. In those early days, bees were exploited more than kept. Honey was, no doubt, simply collected from a hollow tree or whatever other space the bees had found to build their nest. When bees were kept in containers provided by the beekeeper, these early hives were small, probably by necessity, because in early times tools were limited and large containers hard to come by. Furthermore, because of the beekeeper's limited knowledge of bees and colony management, swarming was encouraged. Collecting swarms was the only method available to increase holdings and to replace winter losses. Small containers would also encourage swarming. Hives were made of whatever was available to the particular culture — pottery, woven material, mud and wattle, or a section of hollow log. Later, the more advanced societies employed skeps (domed hives made of twisted straw).

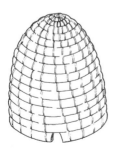

The skep was an immediate precursor of our modern wooden hives.

All of these early hives were crude, at least by our standards. Access was from the bottom or end. No one had worked out a method for removing the comb. The only means known to remove the honey was to cut out and crush the comb. The result was that the colony probably died, either immediately, or later for lack of food.

The philosophy of early beekeeping was simple: have many hives, swarms are good, kill off the good honey producers in the fall in order to take their honey, start over in the spring, encourage those overwintered hives (the weak ones) to swarm. The result — breed to a weakness, breed to swarm.

Very little was known about the bees themselves during this early era, and what was known tended to be localized. The Romans developed some knowledge. They fed the bees in times of need and began to wonder about them, but they had little or no understanding of what went on in the hive. This was at least in part because it was difficult to observe life in a hive without removable comb. These early beekeepers misunderstood or simply did not know about such things as the sexes of the three bees — the worker, the drone, and the queen. They did observe enough to know that there was a monarch, but thought the queen to be male, or king. The facts of mating were vague; one belief was that after eggs were laid in the cells, sperm was then somehow spread over them by the drones.

The existence of wax was recognized, but it was thought to come from outside the hive, probably collected from plants; pollen pellets carried in by the bees could easily be mistaken for wax. There was no understanding of pollination in these early times, and so, no recognition of the need for it or of the bees' contribution to the process.

During the Middle Ages, protective clothing for beekeepers was devised. Up until this time, personal protection was minimal, often no more than a cloth of some sort thrown over the head.

The Second Era

The second era of beekeeping began around 1500 and continued until about 1850. During this time, with the expanding knowledge of science in general, knowledge of beekeeping also increased. Beekeepers were learning about what went on inside the hive and, consequently, about bee biology. Better beekeeping methods and equipment allowed more observation and understanding. During the 1500s, the queen was recognized as an egg-laying female. During the 1600s, drones were recognized as males, although there was still little understanding about mating. During the 1700s, several facts became known: the workers were recognized as females, the source of beeswax became known, the source of pollen and the bee's role in pollination became understood, and the primary facts about mating were recognized.

Honey bees are not native to the Western Hemisphere. They were first brought to North America during this second era, arriving soon after the Pilgrims. No actual account of the first bee importation exists, but they apparently first arrived on the East Coast around 1628; by 1850 they had reached the West Coast, carried there on ships.

Through 1850, honey was most often taken from the hives by destroying colonies. Many methods were devised to take the honey without killing bees, but none was truly successful. The beekeepers' constant goal was more control, and they tried many kinds of hives and frames, including some movable frames, but the bees continued to fasten comb to the sides of the hive. The idea of movable frames was correct, but no one came up with just the right configuration. They were missing some essential understanding.

The Third Era

During the early part of the third era, a basic concept of modern beekeeping was discovered. It revolutionized beekeeping and is still valid today. This concept, the existence of bee space, was discovered or realized in 1850 by the Reverend Lorenzo L. Langstroth and resulted in the Langstroth hive. The design and dimensions of the hives and the movable frames used today are based on Langstroth's discovery, and Langstroth has become known as the father of modern beekeeping.

Bee space is the space within a hive that the bees will recognize as inviolate. It is space the bees leave open to allow passage within the hive. In it they will not build comb or deposit propolis (a resinous material collected from the bark and buds of certain trees). Bee space is generally considered to be ⅜ inch wide, although in practice it can be more and is often less. Natural nests reflect bee space in the spacing between adjacent combs. Manufactured equipment recognizes bee space in the spacing between adjacent frames, between the end bars of frames and the walls of the hive, between the bottom bars of one set of frames and the top bars of the frames immediately below, and so on.

Bees do not completely respect bee space, especially at the height of the season when storage space is at a premium. You will find many instances in which comb is built or propolis is deposited in these supposedly inviolate spaces. (It is up to the beekeeper to remind the bees of their responsibilities by periodically scraping these areas clean. Otherwise, the hive parts become thoroughly glued together and efficient hive management is severely inhibited.)

Shortly after the discovery of bee space came the development of several important pieces of equipment — removable frames, beeswax foundation, the honey extractor, and the queen excluder. The connection to bee space can be seen for each of these. Bee space allowed the free use of movable frames, which were not fastened to the walls of the hive by the bees. These frames were used most efficiently if the wax could be renewed periodically, using foundation. A queen excluder kept the queen from laying eggs in the honey supers, and honey extractors could take advantage of the movable, renewable frames.

Together these allowed for greater ease in hive manipulation and greater opportunity to study the workings of the hive. Furthermore, there was no longer a need to destroy colonies to take the crop, so the resulting crops were larger. Fewer but larger hives could be kept as well. Langstroth's discovery truly revolutionized beekeeping.

As the third era progressed, beekeeping was mostly small scale — one or two hives to provide for a family's needs, although a few individuals kept enough hives to acquire a little pocket money from honey sales. Langstroth's discovery and the developments that followed gave rise to true commercial operations — honey by the ton — and this happened. Before the 1800s were over, several beekeepers in different parts of the country were becoming full-time commercial operators, some of whom had several hundred hives. Honey was shipped by the carload.

The increased attention to bees and beekeeping as this era moved into the twentieth century had its concurrent increase in academic and government interest. When larger crops of honey became common, and as bees were used more often for commercial crop pollination, beekeeping was more readily recognized as an agricultural component. Federal research facilities developed, colleges and universities offered coursework and degrees, and more attention was paid to the science of beekeeping. Our knowledge and understanding of apiculture has increased immensely, and these advances continue.

The Current Era

As a result of these improvements, the comparative ease with which bees could be kept that resulted from Langstroth's discovery, and the increased emphasis on commercial operations, our current informal classification of beekeepers has slowly evolved. This was all hastened and enhanced by the onset of the Industrial Revolution, which changed attitudes and lifestyles over the years. The three types of beekeepers we generally recognize today have evolved from these changes — hobbyists or backyarders, sideliners, and commercial operators. The definitions of each type and the numbers of hives associated with each one can be different, depending on whose opinion you solicit but, in general, the following definitions pertain.

A hobbyist or backyard beekeeper is one who has one or two (or sometimes as many as fifteen or twenty) hives, kept for any of several reasons, but not primarily for income. Probably 90 percent to 95 percent of the beekeepers in this country can be considered hobbyists.

A sideline operator keeps bees as a small sideline business with the intent of making money. It is supplementary to other full-time employment. He or she may have as few as twelve or fifteen hives or as many as 100 to 200. Most sideliners are in honey production or crop pollination; they sometimes evolve to become commercial operators. Sideliners comprise at least 5 percent of the total.

A commercial beekeeper is one who makes her or his entire income from bees, whether it be from honey production, crop pollination, queen or package production, or some combination of these. A commercial operator has anywhere from a few hundred hives to several thousand. Although commercial operators own the large majority of the hives in this country, they probably comprise no more than 1 percent of the beekeepers.

The beekeeping industry today has other components as well. In addition to those who work directly with the bees — honey producers, pollinators, queen and package producers — there are manufacturers and distributors of beekeeping equipment and supplies; federal, state, and academic research and educational facilities; and publishers of beekeeping magazines, journals, and books. Each of these has its own impact on the beekeeping industry.

Presumably, most readers of this book will at least initially be classified as hobbyist or backyard beekeepers. Who knows where that may lead?

How to Get Started

Many adjectives can be used to describe honeybees — extraordinary, remarkable, amazing, astounding — and keeping them can be a thoroughly rewarding and enjoyable endeavor. However, actually getting started with them is not always easy. With proper direction, the path can be simple and straightforward, but it may also be strewn with difficulties. The following steps will help ease the way.

✖ Start with new equipment of standard design and dimensions. Used equipment and homemade equipment have the potential to create problems that the novice is not equipped to recognize or to handle. If you have at least one set of new standard equipment, you have a basis for comparison. Later, you may choose to build your own or perhaps take advantage of a good deal in used equipment.

✖ Do not experiment during your first year or two. Learn

and use basic methods. Master them. Again, you will have a basis for comparison if you choose to experiment in future years.

✖ Before buying a so-called beginner's outfit, know the use of each piece of equipment and be sure that you need it. Also, be sure that everything is included that you initially need. You may find that you are better off making your own selection of equipment, even though you may spend a couple of dollars more.

✖ Start with Italian bees. They are the standard in this country and the most common and most readily available bees. Those acquired from a competent breeder are gentle and easily handled. In future years you can experiment with other races or strains and, once again, you will have a basis for comparison.

✖ Start with a package of bees or with a nucleus hive (a nuc) rather than with an established colony. Once past the initial awe and apprehension, the novice can easily handle and install a package or a nuc. You will relate to it readily and grow in confidence and competence as this new colony grows in size.

   If at all possible, start with two colonies to give you a basis for comparison. If you have only one colony and it is having a problem, as a beginner, you may not recognize the problem. However, with a second hive for comparison, the shortcomings of the one hive are more apparent. Furthermore, a second colony gives you an invaluable resource, a source of bees or brood, for instance, to strengthen a weaker hive.

✖ Start early enough in the season, but not too early. April and May are appropriate for much of the country, and package bees are readily available during this period, but seek guidance from local beekeepers about timing for your specific area.

HOW TO LEARN ABOUT BEES

- First, try your best to find an experienced, successful beekeeper who is willing to help you. This is not always possible: a good mentor can be hard to find. As you seek out this person, recognize that although it takes several years to gain competence as a beekeeper, having kept bees for several years does not guarantee competence or depth of experience. Some beekeepers have one year of experience many times over. Others, although they may understand bees, may not have the ability or patience to convey information to a beginner. Select your mentor with care. Once found, look over his or her shoulder whenever possible. Ask a lot of questions. Immerse yourself in information about bees before you ever acquire any, so that you may make an informed decision as to whether beekeeping is really for you.

- Read about bees. Many good books have been written about bees and beekeeping, and new ones are appearing regularly. The array can be bewildering. Some books are outstanding, others are not. An experienced beekeeper usually has the basic knowledge to pick and choose, both from the vast number of books available and the contents of individual books. The beginner does not. Appendix A will help you to make an initial selection.

- Join your local beekeepers' organization now, even if you don't have bees. You don't have to own bees to benefit. These groups welcome and encourage beginners and they can help you get started. Beekeeping organizations are a prime source of information about

(Continued on page 21)

books, classes, workshops, and other beekeeping events in the area. Many associations hold annual bee schools for beginners. You can usually get information on your local association from the county extension office or from the beekeeping equipment dealer in your area.

- Subscribe to one of the beekeeping magazines. They are full of valuable information for beginners, are instructional, can lead you to sources, and keep you abreast of developments, both locally and nationally. Too much of importance is going on in the beekeeping world these days for anyone to be out of touch.

✖ Recognize that you may not get a surplus of honey the first year, especially from package bees. The first year is a learning time for the beekeeper and a building time for the new colony.

CHAPTER 2

THE BEES

What Is a Honey Bee?

Beekeepers commonly talk about bees as just that — bees. We usually do not use the qualifying adjective, honey. We know, though, that our bees are honey bees and that they are very specific bees within a very large family. To put honey bees in context, consider that on a worldwide basis there are perhaps 20,000 different species of bees and in North America there are 2,000 to 3,000 of those species. In most of the states we can probably find about 300 species. Of all of these species, worldwide, no more than six are honey bees, and in the Western Hemisphere there is only one species, the Western honey bee. The other five species are native to Southeast Asia.

These are interesting and perhaps impressive numbers, but keep in mind that we are talking only about bees. Wasps, including hornets and yellow jackets, are not included. Bees are not wasps; wasps are not bees. The two are related and have many similarities, but they are also decidedly different. Furthermore, there are more wasps, worldwide and locally, than there are bees. To add to the confusion, in addition to bees and wasps, there are also other insects that look superficially like them — certain flies, for instance. All of this means that on a summer day, when an insect flies by and someone swats at it or ducks away from it and says, "Ugh, a bee," it is probably not a honey bee at all, or even a bee of any kind.

To further clarify the honey bee's place in the insect world, see the chart, which represents an abbreviated family tree for the honey bee. It shows where insects, in general, and honey bees, in particular, fit into the animal kingdom.

Unfortunately, most people do not make a distinction between the various insects of similar appearance. To them, a bee is a bee is a bee. Incidents do occur, though, and the poor honey bee gets the blame. Frequently, the actual culprit is the yellow jacket, which somewhat resembles the honey bee, hangs around at summer outings, and is known for its repeated stings. Actually, the yellow jacket is a wasp, related to the honey bee only at the suborder level. The bumblebee, a closer relative of the honey bee, adds further to the confusion because it really *is* a bee, (although not a honey bee), is found in many of the same places as the honey bee, and looks a little like a honey bee.

Let's get back to our honey bee now. It is a very specific bee in the world of bees, known scientifically as *Apis mellifera* and known commonly as the Western or European honey bee. The honey bee's origins are thought to be the Near East, from where it spread naturally into Africa, Europe, and Asia. It is now found in most of the rest of the world, having been carried there by settlers in the early days of colonization.

As *Apis mellifera* has evolved over the millennia, many subspecies or races appeared. These are found in parts of the world where this bee has spread naturally, adapting to the particular climatic and geographical specifics of the different regions.

As you browse through the various literature, you will find reference to a number of different kinds of bees: Italian, Carniolan, Caucasian, Starline, Midnite, Double Hybrid, and Buckfast are examples. Often, little information describing these different bees is included. I'll try to clarify this a little.

First, a reminder. All of the honey bees available to us in this country are of the same species, *Apis mellifera*, the Western honey bee. Other species of honey bees do exist in the world, but all of the honey bees normally kept or found in North America and South America, Europe, Africa, and Australia are of this same species. The other species of honey bee are found primarily in Southeast Asia and include *Apis cerana*, the Eastern honey bee; *Apis dorsata*, the

FAMILY TREE FOR THE HONEY BEE

Kingdom	Animal
Phylum	*Arthropoda:* insects, spiders, and crustaceans
Class	*Hexapoda* (six-footed): insects, about 88,000 North American species, and perhaps 1 million species worldwide
Order	*Hymenoptera* (membrane-winged): bees, wasps, ants, sawflies, and horntails, about 17,000 North American species
Suborder	*Apocrita* (referring to a constriction in the abdomen): bees, wasps, and ants, about 16,000 North American species
Superfamily	*Apoidea:* honey bees, bumblebees, and orchid bees
Family	*Apidae:* bees, about 3,500 North American species
Genus	*Apis:* honey bees, six species worldwide
Species	*mellifera:* the European or Western honey bee, the one species in the Western Hemisphere

giant honey bee; *Apis florea*, the little honey bee; and at least two more recently acknowledged species, *Apis koschevnikovi* and *Apis andreniformis.*

Going back to *Apis mellifera*, many races of this species exist throughout the world. Some of the best known are the Italian (*Apis mellifera ligustica*), the Carniolan (*A. m. carnica*), the Caucasian (*A. m. caucasica*), and the African (*A. m. scutellata*). Because all of these races are of the same species, they can crossbreed. From this crossbreeding we get the hybrids — Buckfast, Starline, Midnite, Double Hybrid, and Africanized, for instance.

Races are relatively easy to understand and they are discussed often enough so that we have a sense of their background and origin. For the most part, they are the result of evolution in geographic isolation (Italians on the Italian peninsula, for instance) where the specific climate and vegetation influenced their development over the ages. Each race has specific traits that relate to the geographic origin of that race.

Hybrids carry things a step further. Rather than letting nature continue to take its course, we have stepped in and crossbred the races, attempting to develop a "better" bee more quickly. The specifics of this betterment are a little different for each of the several hybrids available today, but they commonly include gentleness, increased honey production, and suitability to particular environments or conditions. Sometimes, this all works out and we have a bee that is better suited for a particular situation — the Buckfast bee for British-type conditions, for instance. Other times, it doesn't work out as well — the Africanized bee is an example. (This latter, of course, was not a deliberate introduction.)

The Midnite hybrid is derived from Caucasian and Carniolan stock. The workers are dark in color and known to be extremely gentle. They are said to work at lower temperatures than other bees and to have excellent honey production capabilities. The Starline hybrid comes from Italian stock. The bees are yellow with dark stripes and markings. Their stated desirable traits include good honey production, rapid buildup in the spring, and gentleness. The Double Hybrid is a cross of the Starline and the Midnite lines and, presumably, combines the better characteristics of both these lines. The Buckfast hybrid, the result of many years of effort by Brother Adam, a monk at the Buckfast abbey in Great Britain, derives primarily from a cross of native British black bees (*Apis mellifera mellifera*) with Italian bees (*A. m. ligustica*). The Buckfast is reported to have superior honey production, gentleness, a compact brood nest, and the ability to winter on limited stores.

Maintaining a line of hybrid bees is not simple for a breeder. Many colonies are involved and constant evaluation and reevaluation of the line are required. Breeding stock is maintained through artificial insemination. The procedure is expensive in time and in money. For the individual beekeeper, maintaining a line is easier: it's a matter of buying a new queen periodically.

Although all of the above-named races and hybrids are found in this country, the Italian bee is by far the most common race. It is an all-around good bee for our purposes and was first imported into this country in the mid-1800s. The Italian bee, or stocks that are derived from it, now comprise the majority of bees in this country, both feral and kept. This, means of course, that the same proportion of drones in this country are of Italian stock. A random queen flying out to mate is most likely to encounter and to mate with Italian drones. Therefore, any colony headed by a hybrid queen should be requeened at least every 2 to 3 years, although more often is probably better to ensure that the line is not lost through crossbreeding during supersedure or swarming.

Occupants of the Hive

Three kinds or castes of bee are found in a hive of honey bees: the workers, the drones, and the queen. Each is essential to the life of the colony, and each has a specific role and specific duties. Each is indispensable over the long term, although drones are not present at all times of the year, and even the queen may be absent under certain circumstances. A minimal number of workers is always present.

The total number of bees in a colony varies over the year, but for an average colony in good health at the peak of population, we can expect to find one queen, at least several hundred drones, and as many as 60,000 workers. There are exceptions to these numbers, but we will assume for the moment that they are more or less firm.

The Worker   The Queen   The Drone

All three kinds of bees are, of course, honey bees, so the obvious question is: How and why are they different? The first distinction is their sex. The queen and workers are female; the drones are male. The second distinction is the work or contribution they each make to the life of the colony. The queen and the drone have one basic contribution — reproduction. The queen's duty is to lay eggs; the drone's is to mate with a queen. The workers do the rest. They carry out the many and varied tasks essential to the life of the colony; that is, they work. Let's look at all three kinds of bees more closely.

The Queen

The queen is the most important bee in the colony. Her importance derives from the fact that she is unique. She is the only bee in the colony capable of laying the fertile eggs that grow and produce the life force of the colony — the workers. Those same fertile eggs also produce queens, should any be needed. A further part of the queen's uniqueness is that she is the mother of all of the bees of the colony — workers, drones, and any queens the colony may choose to raise.

The queen's life span is relatively long for a bee. It can be measured in years; 2 or 3 years is perhaps normal, although her life span can be longer and often is considerably shorter.

By her appearance, she is obviously related to the other bees of the colony, but there are differences. She is the largest bee, having a long, tapering abdomen, usually without color bands. Her wings, proportionately, are short, and there are no pollen baskets on her hind legs. A less obvious difference is that the queen has no wax glands for the production of beeswax. She does have a stinger, although she uses it very selectively — for stinging other queens. Reports of a queen stinging a human are very rare, almost nonexistent. Apparently, if a queen does sting a human, it is accidental or the result of extreme provocation.

The queen has two principal activities. She lays eggs, normally 1,000 to 1,500 per day, and sometimes as many as 2,000. Furthermore, by her presence she maintains colony morale and cohesion. A brief life history of the queen is as follows: An egg is laid in a suitable cell; the egg hatches to become a larva; the larva is fed royal jelly (bee milk) for the full duration of its larval life, then spins a cocoon and becomes a pupa. The pupa matures and emerges as an adult queen, which after a few days of maturation goes on a series of mating flights, mating with a dozen or so drones during a 1- or 2-day period. She then returns to the nest carrying a lifetime supply of sperm from these matings. In a few more days she takes up her duties, laying eggs. She may never again leave the colony.

The Drone

The drone, although shorter than the queen, is bulkier and noticeably larger than the workers. His life span is variable but rarely longer than a few weeks. Drones do not normally live through the winter. Their period of existence is what we might refer to as the active season of the colony — the part of the year when the colony is growing, plants are blooming, bees are foraging, and queens may be flying out to mate. Drones are born throughout this active season. The life span of an individual drone is tied to the exact time in the season when he is born, and to his success or failure in fulfilling his role — mating with a queen. The end of a drone's life is often abrupt: if he successfully mates with a queen, he dies in the act. If he is not successful and has not died of old age, he is evicted from the colony at the season's end to die of starvation.

Other than mating with a queen, a drone does no work. He is, in fact, not equipped to work. He has no wax glands, no pollen baskets, no stinger, and no capability to collect nectar or pollen. In the usual course of events, drones are not seen outside of the hive. Their only reason to leave the hive is to seek out queens with which to mate. Because mating takes place high in the air, we seldom see drones (or queens) except in the hive or at the hive entrance as they depart and return.

The Worker

The worker is the smallest of the three bees in the colony and has the shortest life span (although individual drones may live shorter ones). Her adult life span is probably about 4 to 5 weeks during the summer, the busy time of a bee's life. The length of her life is governed by the amount of work done. She wears out her body with her labors. In the late fall and winter, with little or no work to be done, a worker may live for several months. We will consider the worker's life in some detail as we proceed.

Development

Each honey bee's life can be divided into segments. For the workers, there are three — brood, house bee, and field bee. For the queen and drones there are two — brood and adult, although the adult stage can be broken into two phases, maturation and functional.

Brood Development

For all three castes, life starts as an egg and proceeds through the larval and pupal stages of development before emergence as an adult. The timing of this development varies for the castes. It is shown here in days.

BROOD DEVELOPMENT (IN DAYS)				
	Egg	Larva	Pupa	Total
Queen	3	5 ½	7 ½	16
Worker	3	6	12	21
Drone	3	6 ½	14 ½	24

For each caste, the beginning is the same. In fact, all castes arise from identical eggs. Any egg produced by a queen honey bee has the potential to become another queen, a worker, or a drone. The first differentiation occurs as the egg is laid. The queen has the capability to cause each egg to be fertilized or not to be fertilized as it is laid. She does this by releasing or withholding sperm as the egg passes through the oviduct. A fertile egg, after hatching, may develop as either a queen or a worker, depending on other factors that we will consider shortly. Unfertilized eggs develop into drones. This latter process is known as parthenogenesis, or virgin birth. This means that a drone has a grandfather but no father.

Worker Development

A worker's life is divided into three segments: brood, house bee, and finally, field bee.

Each bee starts as an egg. As part of the act of laying, the queen first checks the cell where she plans to deposit the egg. She wants to be sure that the cell is empty and clean. She then checks to determine whether the size of the cell is that of a worker or of a drone. If the cell is of worker size, approximately ⅕ inch in diameter, she causes the egg to be fertilized as it passes from her body. She does this by releasing sperm from her spermatheca, the organ of her body that contains the sperm received from the multiple matings that took place earlier in her adult life. If the cell is of drone size, approximately ¼ inch, she does not release sperm. After 3 days, the egg hatches to become a larva.

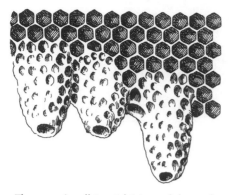

The queen's cell (on right) is much larger than the workers' cells (on left), to accommodate her longer body.

For the duration of her larval life, a young worker is fed continuously. For the first 3 days, the food is exclusively royal jelly, a secretion from glands in the heads of the nurse bees. After 3 days, feeding continues, but the worker now receives ordinary brood food — glandular secretions mixed with honey and pollen.

At the end of this larval period, the developing worker spins a cocoon and enters the pupal stage. For 12 more days the pupa lies quietly and matures, after which it emerges as an adult bee. The total brood development time is 21 days. At maturation and emergence the new young bee is fully grown. She has reached her full adult size, although not all of her muscles and glandular systems are fully functional. This development, from egg through emergence, which is the brood period, takes place within a waxen cell in the heart of the colony.

Soon after emergence, the worker begins her labors. The first half of her adult life, 2 to 3 weeks, is spent in the hive as a house bee. The remainder, about 2 weeks, is spent as a field bee. As adults, these workers perform all of the many and various tasks, other than reproduction, that are necessary to the success of the colony. The actual tasks performed by each worker at any particular time depend on the needs of the colony at the moment and the particular capabilities of the worker at that time. These tasks include such activities as cleaning cells, feeding larvae, building comb, attending the queen, cleaning the hive, processing nectar into honey, guarding, and ultimately, foraging for nectar, pollen, propolis, and water. Nectar and pollen are the food supply and are collected from flowers. Water is used to dilute brood food. Propolis, a resinous material collected from the bark and buds of certain trees, is used to strengthen the comb and to seal up cracks and crevices in the hive against the weather.

Queen Development

The brood stage of the queen's life is similar to that of the worker. However, three things differ: the diet, the timing, and the cell in which she develops. The worker larvae are fed royal jelly for only 3 days, but the queen larvae receive it for their full larval life, which is shorter than that of the worker. The cell for a queen is also much larger than that of either the worker or the drone to accommodate the much longer body of the queen.

On emergence, the queen is not fully developed and functional. She must grow into her job. Initially, she does much the same as workers (and drones) do: she eats, wanders about the brood nest, rests, and completes her muscular and glandular development. About 3 to 5 days after emergence, she makes the first of two or three orientation flights. She must learn her surroundings so that her subsequent mating flights will proceed smoothly. After another 3 to 5 days, she will make her actual mating flights, and another 3 to 5 days later, she will lay her first egg. Her life as a queen has begun.

Once egg laying has commenced, the queen will not leave the hive again in the ordinary course of events.

Drone Development

The drone's development is not significantly different from that of the other two castes. The brood stage is the longest of the three stages, but otherwise is most like that of the workers. When drones emerge, they also stay in the brood area for a certain period while eating, resting, and maturing. It takes about 2 weeks for a drone's reproductive organs to mature. During this time, the drones make orientation flights. As with the queen and the workers, drones must learn their way about the countryside.

Once mature, the drone spends his days alternately flying out to find a queen and eating and resting in the hive. Drone flights take place in the middle of the day and last for perhaps 30 minutes. On one of his many flights, a drone may find a queen and mate, thus ending his life, or he may die one day of old age without ever accomplishing his mission. Hundreds, perhaps thousands, of drones exist for every queen, limiting the possibilities of any particular drone mating. At the end of the active season, any remaining drones are forced from the hive to die. They are seldom allowed to winter over.

Because of his casual and often idle lifestyle, the drone has a poor reputation with many beekeepers, who feel that because drones don't contribute to the hive, get rid of them. Efforts are often made to this end, destroying cells, brood, and sometimes, even the adult drones. This is a short-sighted attitude. Drones are necessary to the well-being of the colony, and workers will go to great lengths to raise replacements during the active season, leading to the expenditure of resources that could be better spent raising workers or foraging.

Activities and Behavior

Honey bees are insects. Beekeepers often lose sight of this fact and attribute to these fascinating creatures a far higher level of intelligence and mental capability than they actually have. For the most part, bees have little control over their actions and their activities. They don't think — they react, and their reactions are in response to specific stimuli. Bees are thought to be genetically preprogrammed from birth, their abilities at any given time being closely tied to physiological development. This theory has been supported by demonstrations showing that workers raised from artificially incubated brood, and having no contact with adult bees, behave in the same manner as workers raised in a normal hive situation.

An individual bee's reaction to stimuli depends on its physiological development, which continues over at least the first half of a bee's adult life. For instance, not every bee can or will sting. The inclination and ability to sting are tied to age and maturation of the bee's venom glands, which are not fully functional until about 2 weeks into a bee's life.

I am alluding here to factors governing a bee's internal organs and systems. These include muscular development, which controls flight; the nervous and sensory systems, which allow for the detection of stimuli; glandular development, which permits wax and venom secretion; and genetic composition, which affects the tendency towards aggression or resistance to disease.

As might be expected, external factors also play an important role. Odor, light, heat and cold, and sound (or more properly perhaps, vibration) — the absence or presence of these factors can be significant in interpreting and controlling activities and behavior.

Time is also a factor in a bee's life, in two respects. One is reaction time. Bees react to stimuli with amazing speed. This is typical of many insects and can be attributed, at least in part, to the very short pathways from the sensory cells to the brain. Bees also have a time sense, and respond to timed activities, such as the availability of food at a particular point and place in each day. Such responses are known to be reactions within the nervous system to a reward of sugar.

Patterns of Behavior

If all of this is true, patterns of behavior should be evident — and are. For instance, a bee returning to the hive with pollen pellets carried in her pollen baskets sometimes loses them before arriving at a storage cell. Even so, she will continue through the series of actions necessary to unload the pellets, not realizing that her baskets are empty. The process of collecting, transporting, and storing pollen is carried through to the bitter end, despite its futility.

Another example of patterned behavior is the bees' reaction to smoke. The judicious application of smoke to a colony brings about a set pattern of responses involving engorgement with honey and, for reasons not completely understood, reduces the defensive reactions of the bees.

Interpretation of Behavior

Honey bees do not have human characteristics or traits. They are not happy, they are not content, they are not mean; they cannot be accurately described by any of the terms that we use to describe human personality or temperament. However, anthropomorphism, the attribution of human traits or characteristics to nonhumans, is a handy way to interpret and to discuss honey bee (and other animal) behavior. Of course, it can lead to inaccurate and improper interpretations, which can be misleading and, sometimes, dangerous. But, as long as the true interpretation is fully understood, anthropomorphism is probably an acceptable practice.

A classic example of anthropomorphism is to judge a particular colony as mean and to accept meanness as its personality. Although some colonies have a threshold of defensiveness that makes them "mean", it could be that a colony is suffering from some outside influence that has temporarily lowered its threshold (constant harassment by a skunk, for instance, or excessive shade). Correction of the problem could restore the colony's equanimity.

Intelligence and Learning

If intelligence is the ability to apply knowledge, to reason, and to think abstractly, bees do not have intelligence. However, they *are* capable of learning because their behavior can be modified by experience. But, the ability to learn is not intelligence. Some examples of bee learning include the association of food and odor, the timing of food availability, and the memorization of the terrain within a fairly large area around their hives.

Do Bees Know People?

Do bees get to know their keeper? The answer to this question probably has more to do with the activities and behavior of the beekeeper than with those of the bees. It seems unlikely that bees get to know any one person. One reason is that the average beekeeper does not visit each hive often enough for the bees to come to know her or him. A bee lives perhaps 4 to 5 weeks during the active season. The average beekeeper probably visits each hive no more than once every 2 weeks or so, and many visits do not involve a detailed inspection of the hive contents. Most of the bees don't have any real exposure to the beekeeper.

Bees may react differently to different people. Sometimes it's chemistry. Bees' lives are governed by pheromones, or odors. Humans have odors that are not noticeable to us but are noticeable to other animals, including bees. The odor of some individuals is probably offensive to bees.

Aside from odors, there are people who get along with bees better than others do, but not because the bees "know " them. It has more to do with the beekeeper's attitude and understanding of the

The most important aspect of understanding the activities and behavior of bees, whether it be an individual bee or an entire colony, is to recognize that just about every bee action is attributable to some kind of situation or stimulus. The more that a beekeeper understands why a bee does what it does, the better beekeeper he or she will be. Take nothing for granted. Observe carefully and question at length. Activities and behavior of honey bees is one of the most fruitful areas for study as you develop as a beekeeper.

bees. If a beekeeper approaches a hive with confidence, opens it carefully with a minimum of thumping and banging, and does not crush or otherwise agitate the bees, they tend to remain calm. If the beekeeper is nervous or apprehensive, this can be communicated to the bees through his or her actions, and they are likely to react.

Life of a Worker Bee

When we think about bees and life in the hive, we think primarily of the worker bee. Although the drone's role in the total scheme of things is critical, we don't think of him on a daily basis, as we do the workers. And, although it isn't quite this simple, you might say that once mating is over and egg laying has become routine, we can forget about the queen unless a problem arises. For most other activities, the worker is the primary life force of the colony, whether it be brood rearing, foraging, honey production, or any of the other tasks that take place within and without the hive. A closer look at the worker's many duties and how they are performed is worthwhile.

Any worker bee is capable at some time in her life of carrying out any task within the colony, except those that are exclusive to the queen and the drones. This statement can be qualified by recognizing that there is a genetic factor that influences capability or inclination to do certain tasks, but for our purposes this factor is not important to understanding the motivation and actions of the workers.

A worker bee emerges from her natal cell as an adult, ready to work almost immediately. She is full size at emergence, although certain of her glands and muscles must develop and mature. After an initial brief period of eating and grooming, this new, young bee begins to work. Her first duties are performed in the heart of the brood nest, and she does not stray far for several days. However, as time passes, this bee undertakes a range of duties, each dependent on at least three factors: her age and corresponding capabilities, the needs of the colony, and the time of year.

Life inside the hive is much more varied than life outside it, but over the course of her life span, a worker spends more or less equal amounts of time as a house bee and as a field bee. Let's look a little more closely at both of these phases.

The House Bee

The young bee is born in the heart of the hive, in the brood nest, where it is dark and warm. She remains here for at least several days, but eventually her work takes her slowly toward the hive entrance. In fact, during her days as a house bee, she makes periodic forays to the hive entrance and beyond to take orientation flights. These forays serve a dual purpose. They allow her to learn where she lives so that she can find her way home when she goes out on her regular trips as a field bee, and they allow her to exercise and strengthen her wing muscles.

Actually, during this time, and even later as a field bee, our bee is not "busy as a bee." In a sense, she is no busier than any other animal. About one-third of her time is spent resting or sleeping and another third may be spent patrolling, moving through the hive looking for work. If, during her patrolling, she comes across a job that needs to be done, and it is within her capabilities at that moment in her life, she will undertake it. She does not concentrate exclusively on any one task in a given time period. She is usually capable of undertaking more than one task during any developmental stage and will take on whichever of these she encounters.

TASKS OF A WORKER BEE

As a young bee, her abilities are limited. At emergence, her muscles are not well enough developed to enable her to fly. Many of her glands are not yet functional, limiting her ability to perform certain tasks (wax secretion, for instance). With the passing of time, these muscles and glands mature and she undertakes the various needs of the hive in a more or less set order. An abbreviated list of these tasks follows. It shows the approximate order in which she is able to perform them, progressing from the heart of the brood nest to, ultimately, the great outdoors.

- Cleaning brood cells
- Attending queen
- Feeding brood
- Capping cells
- Packing pollen
- Secreting wax
- General cleaning of hive
- Receiving and processing nectar
- Guarding
- Foraging

Her abilities are not cumulative and she cannot perform certain tasks continually. For instance, very early in her life, each worker bee's hypopharyngeal and mandibular glands begin to secrete brood food. These glands function for this purpose for about 1 to 2 weeks. She is then no longer able to feed brood. During the period that these glands are functioning, they do not secrete food continuously, so a given bee spends very little time actually feeding brood. Between bouts of feeding, which are often quite brief, she performs other tasks.

Not every bee performs every role for which she is qualified. There are only so many cells to be built, so many larvae to be fed, and so much guarding to be done, even in a busy season. Furthermore, many tasks are strictly seasonal. For instance, little or no foraging, nectar processing, or comb building is done during the winter months in most of the country, and brood rearing is cut back drastically or stopped altogether.

Finally, at some point, the needs of the colony for new food gatherers, the ever-changing capabilities of her body, and the pressures of the new generations developing behind her push her out the door and into the life of a forager.

The Field Bee

When the house bee graduates, she becomes a field bee and will spend the rest of her days foraging. She may collect nectar, pollen, water, or propolis during this period. However, she will tend to concentrate on one or two of these.

Foraging is not random. Except perhaps for a few scout bees, each bee, as she leaves the hive, goes after one of the specific raw materials of colony life. She collects this same material exclusively on a given trip, perhaps all day, and in some instances, throughout her entire time as a field bee. This constancy is one of the reasons that the honey bee is so valuable as a pollinator. She does not go from one flower species to another as she forages for nectar or pollen, but stays with the same species for at least that trip and usually for many trips. The pollen is then moved from like-flower to like-flower, and pollination is accomplished.

Many variables affect the length of a bee's life. Her labors, both in and out of the hive, take a toll on her body. One visible manifestation of this is in the wear on her wings. This comes partly from brushing and rubbing against other bees in the busy hive and from contact with flowers and foliage as she forages. Eventually, she can no longer fly while carrying a load. Her life is effectively over and she will soon die. Actually, the bee's worn wings more or less coincide with a particular physiological development: a bee is capable of digesting only a given amount of sugar in her life, and once

The field bee's wings are worn down by repeated contact with flowers, foliage, and other bees in the hive. When it can no longer fly, the bee soon dies.

she has processed that amount, her life ends. During the active season when the bees are working harder, life ends sooner. In the off season, the bees tend to be idle and eat less, so they live longer.

The Needs of the Colony

We have alluded several times to the needs of the colony. In a colony with little brood to raise at a particular time, young bees pass quickly through that phase of development to the next — wax secretion, perhaps. The capability to secrete brood food disappears or atrophies. This is understandable and reasonable.

However, there is another aspect to the way a bee responds to needs of the colony. Sometimes a massive disruption in colony life takes place. Perhaps insecticide has wiped out all or most of the foraging force. House bees will then mature very quickly. Some of the older house bees will graduate early and take up foraging. Younger house bees will develop faster to fill in for those departing older ones. The overall balance of activities in the colony will be quickly restored, although in numbers smaller than before the disruption.

Similar changes can take place if an insecticide is inadvertently brought back to the hive by pollen foragers. Stored in the pollen cells, insecticides may cause the death of large numbers of newly emerged young bees as they feed — potential nurse bees. Very quickly, the capability of some of the older bees to secrete brood food will regenerate, thus restoring the balance.

Of course, none of this is a result of a direct recognition of needs through any kind of judgment or thought process on the part of the bees. Everything takes place in response to the absence or the presence of pheromones given off by adults and brood, which signal the status and the needs of life in the hive.

Cycle of the Year

What is the beginning of a bee's year? We have several choices, and in a sense, the answer depends on what aspect of beekeeping we are concentrating on at the moment. Actually, as soon as we start thinking about a bee's year, we tend to focus on hive management, which leads into the beekeeper's year. The bee's year and the beekeeper's year are not necessarily the same. The fact that a nice day in early spring is when we first open the hive has little bearing on what the bees have been doing all along.

A honey bee colony is active year-round. Although there are periods of relative inactivity, there is no hibernation. The type and level of activity in the colony throughout the year are dependent on the season of the year, the climate and geography of the specific locale, and the current weather. Most of us deal with bees in a temperate climate, and this is a governing factor.

Although the colony is potentially active year-round, its level of activity varies considerably, as does the population level. This population level is a reflection of hive activity. When activities are low, as in the winter, population is low. When activity level is high or has the potential to be high, population is high. For much of North America, the population cycle of a colony is represented in the chart provided here, perhaps with local variations.

ANNUAL POPULATION CYCLE OF A COLONY

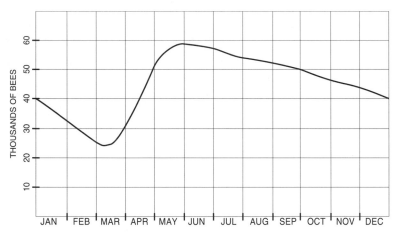

The curve reflects a healthy, normal colony with adequate food reserves and with normal forage available. Notice that the population may fall off 50 percent or more over the year. A curve representing a specific colony in a specific location would show more short-term variability, reflecting the actual conditions affecting that colony.

The curve as shown is a representation of colony life for an existing colony, one that has wintered over for at least 1 year. We will arbitrarily use mid-March as the beginning of its new year. The reason is obvious. This is the low point in a colony's population, the time when the increased stirrings of new life offset the winter decline, and the population count is rising after a winter's rest.

There are other dates that could be chosen as the start of a new year. Swarming time is one: it represents a new beginning for both the swarm and for the parent colony that cast that swarm. The installation date of a new package or of a nucleus is another. However, these are best understood if we first outline the rising and falling of life in an established colony.

The Cycle

As winter passes, the bees cluster, keeping each other warm, while normal colony life goes on inside the cluster — eating, sleeping, household chores, raising brood, and so on. The bees even fly out if the weather permits. However, bees die every day, and as a result, population falls.

Brood rearing has been underway, perhaps since mid or early winter. The exact time it begins in the new cycle is variable, depending on such factors as the climate, the race of honey bees, and the conditions in the colony. If all goes well, the rate at which new bees emerge exceeds the rate of die-off, and the net population increases. With the stimuli of spring — lengthening days, warmer weather, blooming plants — the rate of population growth increases, leading up to maximum numbers at about the time of the summer solstice.

After the solstice, the days begin to shorten and the bees take this as a signal to begin cutting back. Up to this point, colony activities have concentrated on population expansion and nectar and pollen collection. Now, emphasis ever so subtly shifts to preparation for winter. Brood rearing continues at a good rate, but by paying careful attention you can see that it is slowly tapering off. In the early season, the bees collected great quantities of pollen and of nectar, but the bulk of these were consumed in brood rearing. Now, stores are beginning to build up and the beekeepers' honey production is in full swing. Of course, the bees consider it their honey for the winter ahead. That's fine — there should be plenty for all.

With late summer and the onset of autumn, population drops further and activities slow. Brood rearing cuts back more and life in the hive moves at a less frenetic pace. As the nights begin to cool (and later, the days), the bees spend more time in their cluster, and brood rearing slows to accommodate this change. No more brood can be raised, at any time of year, than can be covered and kept warm by the adult population as they cluster. No matter what the season, the brood nest must be maintained at approximately 94°F (34°C). Sometime in the late fall (the exact time is dependent on latitude and other factors pertaining to the specific colony) the queen stops laying. A 3-week period follows during which the last of her eggs hatch, the brood matures, and then all brood rearing ends. Some weeks later, at least partially in response to the lengthening days, the queen starts laying again and the new cycle is underway.

Although we hear of the queen laying 1,000, even 2,000 eggs per day, she does not start out at this rate. She works up to it slowly, but by early spring her numbers are well into the hundreds and rising. By late spring she has hit her stride, and with the renewed availability of nectar and pollen, the new year is well underway. The cycle is complete and a new one begins.

Colony Reproduction: Swarming

Swarming is a fact of beekeeping life. It is difficult to keep bees and not have to confront swarming at some point, or even many points in a beekeeping career. It may not be a matter of actually having swarms. With careful management, swarms can be controlled — most of the time. But management for swarm *control* is a part of swarming. It is important, it is often misunderstood, and sometimes, it doesn't work. However, a beginner with a package or a nucleus

WHAT IS SWARMING?

To understand swarming, it is helpful to consider honey bees on two levels — as individuals and as a colony. The life cycle of an individual bee is short and that life is "cheap." Turnover of individual bees in a colony is high. In the course of a normal day, many bees are born and many die. But, the colony has the potential for immortality. In practical terms, this doesn't happen. Colonies die as a result of disease, fire, predators, and other natural and unnatural disasters. So, nature must have a way of replacing colonies, of perpetuating the species. Swarming is nature's solution.

Not every colony swarms every year. Left to its own devices, a colony would probably swarm about once every 2 or 3 years. When a colony does swarm, about one-half of the bees of the parent colony leave, taking the queen with them. This potential new colony goes through a process of finding a suitable home, moves in, establishes itself, and continues life as a new colony. Meanwhile, at the old stand, the remaining bees raise a new queen and continue life as before. Where once there was a single colony, now there are two.

hive probably will not have to cope with swarming during the first season. There are no guarantees, though, so be prepared with some understanding of the process.

The following information is not intended to be a complete explanation of swarming and swarm control. Entire books have been written about the process, and research continues. The intent is to expose you to the problem. Once you are comfortable with your bees, look more deeply into swarming and its control and prevention.

In the bees' scheme of things, swarming is good. However, even for bees it is not always desirable for a colony to swarm. It may not have the strength required to do so successfully, the timing within the season may be poor, and the quality of the season may not allow a swarm to survive. However, the colony does not know this. It simply reacts to signals — to factors such as nectar flow, numerical strength of the colony, age of the queen, and other factors that may not relate to the ability of the swarm or the parent colony to survive. Don't assume that the bees know what they are doing.

When Do Colonies Swarm?

Swarming has a particular season, actually two particular seasons. The primary swarm season is in late spring and early summer and lasts for about 6 weeks. The season starts earlier in the south and later in the north. Depending on location, the season may begin during April or May and continue into June or July. It is somewhat variable from year to year, reflecting the actual weather of the particular year. During this primary swarm season, about 80 percent of the swarms of that year occur. You can learn the dates of the approximate swarm season for your area by talking with local beekeepers.

A secondary swarm season also exists that is less well known to many beekeepers. This period extends for about a month, starting sometime in August. During this period the other 20 percent of swarms occur. Swarms from the primary season stand a reasonable chance of surviving into the following year, but long-term survival for these late swarms is much less likely. They do not have time to become reestablished for winter. In addition, the parent colony has been badly weakened and will also have difficulty surviving.

The Process of Swarming

The general reason for swarming is propagation of the species. Why a particular colony swarms at a given time is less clear, but can usually be tied to population and perceived congestion in the hive. The age of the queen is a related factor. Other factors also bear on this process and it is difficult and dangerous to oversimplify any explanation.

The preparation for swarming begins well in advance of the actual day of the event. It is triggered by some combination of conditions in and out of the hive involving population, size of the hive, age of the queen, nectar flows, weather, and other factors. The exact conditions probably vary from colony to colony. It is often difficult to look at a colony that has just swarmed or is about to swarm and discern its precise reason for swarming. Many people believe that it can be explained in simple terms and controlled by simple manipulations. But swarm control requires a thorough understanding of the process and application of specific hive management techniques.

Let's assume that conditions in our colony are such that it is ready to swarm. Some days before the actual event, the queen begins to reduce her rate of egg laying. This is not a voluntary act on her part. It is the result of less stimulation and feeding of the queen by the workers. The queen is not in charge.

About 3 days before a swarm, the queen stops laying entirely, and some of the workers begin to harass her, shaking and mauling her. Presumably, their goal is to cause her to lose weight and bulk, getting her down to flying trim. A queen who is in full egg-laying condition cannot fly far or well and swarms often travel great distances.

Meanwhile, the workers have begun to rear several new queens, one of which will head up the parent colony after the swarm has departed. The swarm actually leaves the parent colony about the day that the first of these queen cells is capped, weather permitting. About 9 days after the swarm has departed, the new queen emerges as an adult to begin preparation for taking over the egg-laying responsibilities for the parent colony. The swarm moves on to a new home, usually some distance from the parent colony, where they establish themselves and begin life as a completely new and separate colony.

From a beekeeper's point of view, swarming is a highly undesirable occurrence. The parent colony has lost about half of its population at the height of the honey production season. It has lost the services of a laying queen for up to 4 weeks while the new queen develops, matures, and begins to lay, which also adversely affects population. Honey production for the season can be severely reduced, if not lost altogether, and the colony may be so weakened it cannot adequately prepare for the coming winter.

Swarm Control

Swarming is controlled by an impulse, a compelling force that takes over the colony some days before the actual emergence of the swarm. That impulse must be satisfied, which happens when the colony actually swarms or when they are made to think they have swarmed. Swarm control and prevention measures that do not recognize and take into account this swarm impulse rarely work. In fact, such measures may do more harm than good to the colony in the long run. Such inappropriate steps include confining the queen, blindly destroying queen cells, and simply adding more supers to the hive to relieve congestion without paying close attention to the overall needs of the colony.

Recognition of the importance of the swarm impulse is one of the most important lessons to be learned in beekeeping. Of course, recognizing the swarm impulse does not in itself control swarming — there are steps that must be taken. The best swarm control measures begin well in advance of the swarm season and are manipulations that prevent the swarm impulse from developing. These include regular requeening, recognizing and relieving congestion, knowing the environmental conditions that encourage swarming (for instance, a warm winter, an early spring, or strong nectar flows), and removing adult bees or brood, either permanently or temporarily.

A new colony started in the spring is not likely to swarm during that first season. A novice beekeeper should not be overly concerned about swarming, and especially should not be carried away with poorly understood or unnecessary swarm prevention techniques. However, this is one of the important areas for the new beekeeper to read about and gain additional knowledge of before the next season. Furthermore, the focus of this reading should be on activities and behavior, and on bee biology. To correctly understand swarm control and prevention, you must first understand the bee.

Nutrition and Feeding

Nutrition is as important to honey bees as it is to any living thing, and yet, bees are more restricted in their food sources than many other creatures. During certain times of year, no outside sources of food are available to them and bees are left to their own resources. They work very hard to build up food reserves for these times. We often lose sight of the fact that bees are storing food for their own benefit, not for ours. They are anticipating the cold rainy stretches when they cannot leave the hive, the droughty conditions when no nectar or pollen is available, or the long cold winter ahead. We, too, must be aware of all of this and do our part to ensure that the colony always has adequate food.

Nutrition

The two principal components of an adult honey bee's diet are honey, which they produce from nectar, and pollen. From these two sources, supplemented by water, the bees derive all of the elements of a balanced diet — carbohydrates, protein, fats, vitamins, and minerals. Carbohydrates come from the honey, in the form of sugar; protein and fats come from the pollen; and vitamins and minerals come from both.

Honey bee brood also requires food during the larval stage. Their first food is bee milk or royal jelly, a secretion from glands in the worker bees' heads. Later in larval life, honey and pollen are added to that diet.

Not only must the bee's diet be balanced, but food must be available continually to stimulate colony activities. The bees are aware of how much food is in reserve and how much is coming into the hive. They adjust their daily activities in accordance with this volume. Bees cannot store energy in their bodies. If there are no food reserves and no nectar coming in, the colony will soon perish. An individual bee can live only a matter of hours without sugar.

Pollen is also critical. Without pollen stores on hand, or pollen coming into the hive, no brood rearing takes place — brood rearing that is underway stops, no new brood rearing will commence, and brood that is developing may be removed by adult bees.

Let's look a little more closely at these dietary components.

Nectar (carbohydrates). Nectar is a watery secretion available from the flowers of many plants. A principal component of nectar is sucrose, a sugar, and it is for this sugar that bees collect the nectar. The bees convert nectar to honey by reducing its moisture content and inverting the sucrose to yield two different sugars, glucose and fructose. Bees can and probably do eat a certain amount of nectar directly, but their main source of energy throughout their lives is the honey they have produced and stored. A colony cannot survive without nectar or honey continually available.

Bees cannot collect nectar from all flowers. Some flowers do not secrete nectar, whereas others do not have enough concentration of sugar to allow efficient collection. They would expend more energy collecting from these latter flowers than they would gain from the resulting honey.

Pollen (protein). Protein, in the form of pollen, is necessary for the development of the bees' muscles, glands, and tissues. It is a component in the diet of the brood and is eaten by newly emerged adults. This consumption diminishes as the adults grow older and stops by the time they reach field bee status. However, pollen is found naturally in honey, so the bees consume some amount of it throughout their lives.

Pollen is the male reproductive part of a flower. As collected by bees, it is a microscopically fine dust that they consolidate into pellets by adding small amounts of moisture from their mouthparts. These pellets are transported back to the hive adhering to the pollen baskets on the bees' hind legs.

Pollen is collected from a variety of different flowers, although not necessarily from the same ones as those from which they get nectar. Bees collect both nectar and pollen from some flower species, while they collect only one or the other from others. Pollens from different plants differ in nutritional content, and therefore, in their value to the bees. They should have a balanced pollen diet, although the beekeeper has little control over this. Many parts of the country have pollen shortages or deficiencies, but bees do not select for content or balance: they just collect what is there.

Pollen must be collected by the bees for it to be valuable to them. In other words, we cannot collect it from flowers and feed it to them. While collecting and storing pollen, bees add enzymes and perhaps other substances that help to preserve it and enhance its dietary value. Old pollen loses potency. After a year, it is of questionable value, but bees do not recognize this. They use whatever is available.

Vitamins. These are essential for the growth and the development of any living organism. Pollen has an exceptionally high vitamin content. Small amounts of vitamins are also found in nectar, and consequently, in honey.

Minerals. These are collected routinely as a part of nectar, pollen, and water. Bees need them, as does any living thing, but different elements and different minerals are necessary to particular functions at specific ages. Both deficiencies or the presence of certain minerals can be detrimental. For instance, salt reduces longevity.

Fats. Little is known about the need for fats. For day-to-day use, fat is available from pollen. There are fat bodies in the body cavity of honey bees that function as production and storage sites for food reserves, but these are developed primarily from sugar, not from pollen. The contents of the fat bodies vary with the season and the age of the bee.

Wax secretion is dependent on sugar consumption, and not, as might be expected, on pollen consumption.

Feeding

Feeding a colony is occasionally necessary and a few words about what to feed are in order here. (When to feed is discussed in Chapters 3 and 4.) Two substances are commonly fed to bees — sugar and some form of pollen or pollen substitute.

Sugar. Two forms of sugar, other than honey, are acceptable for feeding bees: granulated sugar (sucrose) and high-fructose corn syrup. Both of these, when pure, create no problems for the bees. However, both of these (and several other forms of sugar) are found in many "impure" forms that can seriously harm, if not kill, bees. Examples are molasses, candymaking by-products, corn syrup, maple syrup or sap, high-fructose syrup such as surplus from soft drink plants, and others. The only sugar that should be used as feed for bees is table-grade granulated sugar, the same kind you use. High-fructose corn syrup should also be of food quality, with no additives.

Pollen. Protein may be fed in either of two forms — pure pollen or pollen substitute. Pure pollen can be collected at the hive entrance by the beekeeper, then cleaned, dried, and stored. Pollen traps are available for this purpose. It may then be fed back to the bees in times of need. Pollen substitute is a manufactured material containing variously such components as soy flour, brewers yeast, dry skim milk, and powdered casein. Pollen substitute is often combined with pure bee-collected pollen. In this case, the mixture is called pollen supplement. All three — pollen, pollen substitute, and pollen supplement — are fed in the same manner.

Pollen supplements can be purchased through bee supply sources in two different forms. One is dry, simply called pollen substitute. It can be fed as is, inside or outside of the hive, or it can be moistened with syrup and fed as a patty inside the hive. The second form comes in commercially prepared packages, offered under one or more trade names, and is intended to go directly in the hive.

Communications

Communication between honey bees takes place both inside and outside of the hive. This communication is extremely important to the successful life of a colony and it is reasonable to say that honey bees would not exist as we know them if they could not communicate with each other as they do. Examples of the information that bees are able to pass through various means include: locations of specific forage, whether it be nectar, pollen, or water; locations of potential new homes for a swarm; danger warnings; and specific needs within and without the hive.

Information is conveyed through the interpretation of odors (pheromones) given off by bees in particular places and under particular conditions, and through their ability to convey specific information through a so-called dance language. An understanding of the role of pheromones is important to understand the behavior and the motivations of individual bees and of colonies, and although it is not necessary for us to understand the bees' language, it is interesting and at times helpful to be able to do so.

Pheromones

A pheromone is a chemical substance given off by an organism that serves as a stimulus to other organisms of the same species to cause a particular behavioral response. An organism may give off a few or many pheromones under different circumstances, affecting many aspects of its life. Pheromones are usually perceived as odors or tastes. We will simplify pheromones here and think in terms of odors.

A bee's life is largely governed by odors. These odors are associated with most aspects of a colony's activities. For instance, the queen gives off an odor in the hive that signals her presence to the colony at large. The brood gives off odors that signal caste and stages of development and indicate to the workers such needs as those for feeding or for cell capping. Odors deposited in an empty cell by a worker bee can indicate to the queen that the cell is empty and ready for an egg. At the hive entrance, the odor (or the absence of a correct odor) of a strange bee or other intruder indicates to the guards that there is something to be repelled. A homing odor is released at certain times of confusion to help lost bees find their way back to the hive or to help a swarm find its way to a new home. An alarm odor is associated with venom to signal the intruder's location after stinging, so that more stings will be directed there. The list could go on and there are undoubtedly many pheromones associated with a bee's life that are yet to be discovered.

Foraging

Flowers of different plant species vary widely in the quality of their nectar, as measured by sugar content and by volume available per bloom. As a result, these flowers or crops vary widely in their economic value to the colony. It is in the bees' best interest to forage only on those blossoms that yield the richest return. This return is usually determined by sugar concentration, but is also influenced by such factors as amount per individual bloom, concentration of the crop area, and distance from the hive. When bees forage, they do not do so blindly or randomly. Foraging is an efficient undertaking, with the colony exploiting its neighborhood to find the most rewarding sources available. Once an individual bee has found or been directed to a particular source, it remains faithful to that source for an indefinite period. This period is determined by one of two things. Either the source dries up, its time of bloom over, or the bee is made aware of an alternate source that is yielding nectar of greater economic value to the colony.

At a given time, a colony may be happily foraging on a particular source, or perhaps on more than one source if others are available of approximately equal economic value. A scout bee comes across a new source, perhaps something just coming into bloom. The scout believes the new source compares at least favorably, if not better, to the sources currently being worked. It will return to the hive and signal the existence, the location, and the relative value of this new source. But first, how can this scout bee know that the find is worth touting? To understand this, we need to understand what happens to nectar as it arrives in the hive.

The nectar must be processed to convert it into honey. This processing is carried out by the house bees. The foragers, arriving at the hive with a load of nectar, pass it off to the house bees. Each house bee involved in the processing is aware of the quality of the nectar as it comes in. House bees only reluctantly accept loads of nectar from field bees that are of lesser quality than what has been arriving, but eagerly accept loads of higher quality. Thus, field bees are kept apprised of the quality of their contribution by the speed with which they are unloaded.

A scout bee, randomly searching the countryside, may come across a new source that she recognizes to be of exceptional quality, compared to what is currently being brought in to the colony. The scout bee will take a sample of this nectar back to the hive and communicate to hivemates the existence of this new source by dancing.

The Dance Language

Bees have several dances that they use to inform hivemates of availability, quantity, and quality of nectar; honey; pollen; water; propolis; and potential homesites. These are not truly dances but that is a convenient name for these movements.

A bee in the hive dances when it has information to convey. Other bees, potential foragers, are attracted to the movement of the dance and approach the dancing bee. They are able to interpret the dance to learn what is being touted, its quality and quantity, and its location. Keep in mind that bees do not see the dance; it is dark in the hive.

The two primary dances used to signal new foraging locations are the round dance and the wag-tail dance. These are performed by foragers as soon as they return to the hive and take place on the vertical surface of the comb, usually near the entrance of the hive, where potential foragers are resting and waiting. For the purposes of this discussion, we will assume that the new forage source in question is nectar. A forager returns to the hive carrying a sample of this new find. She may also carry on her body odors associated with the flower.

The dance involves quick short steps. The total movement is confined to a relatively small area and may continue for a few seconds or as long as a minute. Other bees become excited by the action of the dance and follow the dancer, antennae toward or touching her. The dancer may be interrupted as other bees solicit samples of the nectar she is carrying, or the dancer may stop entirely and move to a different place on the comb, where she resumes dancing. For each dance, her hivemates are able to gather specific information.

The round dance. This dance is used when sources are within 100 meters of the hive. No other distance information is conveyed. The other bees know only that the new source is somewhere within a 100-meter radius, and as a group they will randomly search in that area. The other information conveyed in the round dance is the type of forage, which is apparent by the samples offered and the odors that may be adhering to the dancer's body.

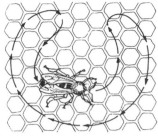

A bee's round dance signals to others the discovery of a new source of nectar or pollen within 100 meters of the hive.

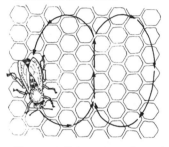

The wag-tail dance — performed in a figure-eight pattern — conveys the distance and direction of a new nectar source that is over 100 meters from the hive.

The wag-tail dance. This dance is associated with sources more than 100 meters from the hive, but the dancer is able to convey more precise information. Within a relatively small margin of error, the exact distance and a direction is indicated.

In the wag-tail dance, the bee moves in a figure-eight pattern, always in the same direction, and the orientation of the figure-eight remains the same throughout the period of the dance. The straight line through the center indicates the direction to the source of nectar, and may be interpreted by its relation to the direction of the sun.

When the forager, the potential dancer, flies to and from the nectar source, she flies on a line that is at a specific angle to the direction of the sun. That angle is reproduced in the hive as the bee dances, using a vertical line on the comb to represent the direction of the sun. As the dancer moves through the dance pattern, she wags her abdomen and gives off low-frequency sound blips. The number of blips, the intensity of the wagging, the number of straight runs, and the duration of the dancing have variously been correlated to the distance to the forage site and its quality. The quantity of food available can also be related to the number of bees dancing.

The dance language of the bees as outlined here is unbelievable, in a sense, considering that a bee has limited brain capacity, no intelligence, and no thinking capability. It becomes even more remarkable when we realize that the bee recognizes the movement of the sun and can adjust the straight-line run of the dance to follow that movement if a large amount of time passes during the dance period. The bee can also indicate the distance that must be flown to compensate for the effects of a crosswind en route, and she will indicate the straight line shortest distance even if the route actually flown must vary to go around or over an obstruction.

The alarm dance. The alarm dance indicates that contaminated or poisonous food has arrived in the hive. The dance starts within minutes of the arrival and is performed by both house and field bees. The bees run in spirals or irregular zigzags, shaking their abdomens vigorously. As a result, all flight activity stops.

Other dances. Other dances have been recognized, among them the sickle or crescent dance, which is a transition between the round and wag-tail dances, and the cleaning dance, a solicitation by the dancer to other nearby bees to clean and groom her.

The Importance of Dancing

What is the importance of the dance language? It is extremely important to the bees and the success of a colony and thus to the beekeeper. But it is not something that we can control or influence in any way. We can only recognize its existence. However, consider the implications. With a dance language, the bees fully exploit the resources of their neighborhood on a timely basis. Information on the comparative quality of sources is gathered and disseminated quickly. Without the dance language, each bee would be on her own, searching randomly through the countryside until she found an acceptable source. Large amounts of time would be spent scouting, and perhaps flying excessive distances, when acceptable forage might actually be located closer to the hive.

Researchers set up a large observation hive in a deciduous forest area. They watched and interpreted the dancing and analyzed foraging behavior of this colony for an entire summer. Over this time, the bees searched a 36-square mile area. They worked relatively few patches each day, but the locations changed almost daily. The conclusion was that the bees kept moving to more desirable patches as the quality changed. They could not have done this without a sophisticated communications system.

The dance language also helps to protect the colony. Foragers subjected to hazards — pesticides, toxic nectar or pollen, predators — either never return to communicate the source, or their behavior is so modified that they do not successfully communicate.

Dancing is probably as important to the long-term success of a colony as the stinging behavior discussed on page 6. Both activities allow colonies to maintain their population levels and, in turn, allow us to enjoy our particular diet and life-style by enhancing the pollination process.

CHAPTER 3

GETTING STARTED

W here will you keep your new hive? For some, the answer is obvious because you have limitations that restrict your choice of spots — a small yard, for instance, or perhaps specific areas of a larger property that must be avoided. Others may have no serious limitations. Let's assume that you are one of the fortunate ones who has an unrestricted choice of locations. What are some of the considerations? And if you are limited in your choice of hive location, how can you compensate for potential problems?

Hive Location

First, what are the bees' preferences? From studies done with swarms seeking new homes, certain preferences are known. Bees favor an elevated location, preferably about 10 feet off the ground. Humans can rarely accommodate them on this, since we must be able to get to the hive ourselves. They also like a southerly exposure. We can usually provide this, but it is not critical if we cannot. Sunlight is also a factor. Bees like a location with a visible lighted entrance, but too much or too little sun is undesirable. In nature, a bee's home is most often found in a hollow tree. Consider the amounts of sun and shade available in such a setting.

Aside from the bees' preferences, we need to consider our comfort and the neighbors' apprehensions, all the while applying a measure of common sense. Let's look at some specifics.

Visibility and Accessibility

First of all, consider that out of sight is out of mind. I don't know of a study done on this, but without question, if your hives are not near home, you'll give them less attention. As a beginner, you want them nearby so you can visit them often. New colonies and new beekeepers need to get together regularly, and it is too easy to put off a visit involving a trip across town — or even just a mile or two down the road. When the hive is in the backyard, it is easy to slip out there at odd moments for a quick look. These casual visits are important, even when you don't open the hive. A hive located at home also saves having to load and unload pieces of equipment every time you work with them. It is annoying to find yourself at a remote site without some essential item of equipment — your smoker or hive tool, for instance. Do your best to place the hives at home or at least nearby.

Vandalism is another problem if your beeyard is remote, and sometimes even in your own backyard. Hives are prime targets for vandals and, occasionally, for thieves. Hives may be pelted with rocks, tipped over or otherwise trashed, or may even disappear entirely. I have had some hives pelted by gunshot, others tipped into a river, and, although I have never lost a whole hive, on a couple of different occasions I have had frames of brood and the queen taken.

The answer to all of this is to make the hives less visible. Use shrubbery, a small section of fence, or even an entire building to mask their presence.

Exposure

Exposure has several facets. First, hives may be mistakenly located where they are overexposed to the elements. Wind and sun can be excessive. If a hive must be placed where there are excessive winds, construct a windbreak. It does not need to be elaborate. Install a simple fence, pile up some hay bales, or plant a shrub or two — anything to break the wind. Be especially mindful of the winter winds. They may not be fresh in your mind when you are setting out bees in the spring, but think back a couple of months. What was this location like in the dead of winter? How much wind was there, how deep was the snow? A hive is not particularly affected by snow piled around it. The snow even gives some protection. However, a location where the hive may become completely covered should be avoided. The bees do need to get in and out on the warmer days in winter.

Next, think about traffic in all of its forms — vehicular, pedestrian, and animal. If the hive is close to, and especially if it is facing, a busy road with its turbulent air currents, bees may be lost as they try to negotiate those currents. Pedestrians and bicyclists will be in the bees' flight path as they come and go, and although this does not mean they will automatically be stung, accidents can happen. Animals, too, must be considered. It has long been known that bees and horses often have difficulty coexisting and should not be kept in close proximity. The odor of horses seems to offend bees and has been known to trigger stinging incidents. This is compounded by the fact that horses are not only less tolerant to stings than other animals, but they react violently, often to the detriment of the horse. Other animals seem better able to take care of themselves, and we seldom hear of serious problems. However, there have been reports of livestock using beehives as scratching posts, often knocking them over. All in all, it is best to give the bees their own space. Skunks and bears are a special problem, as discussed on pages 161-162.

Comfort and Workability

When you are working with your hive, you will of necessity disassemble it. Parts will be strewn about on the ground as you work and you'll also need space to move around a bit. You may want to bring a garden cart or wheelbarrow up close, and perhaps the lawn mower. Be sure all of these will fit in the area you have chosen. Allow at least 3 to 4 feet on all sides of the hive, if possible.

Be careful not to locate a hive on a steep slope. Level ground or a very gentle slope is probably best. A gentle slope is good because it allows cool or moist air to flow down and past the hive and not form a pool around it. If the slope is too steep, you may find covers and other hive parts also flowing down the hill when you set them on the ground. Avoid such steep slopes, as well as depressions where cool or moist air can collect.

Forage Availability

An immediate question that occurs to many potential beekeepers is, "Will my chosen location support a colony of bees?" For the most part, you need not be concerned. One or two hives can be successful almost anywhere on the continent. In fact, some unlikely locales can be very successful — bees are often kept in large cities and in suburbs.

The key to this is the area around you. The normal foraging range for honey bees is 1 to 2 miles and they have been known to travel much greater distances in areas where forage was scarce. The fact that your immediate property has little to sustain the bees is not important. It is its 1- to 2-mile radius that matters. If there is a substantial amount of flowering plants in that radius, you are probably okay.

Of course, if your own property does have good forage, the bees will work it and benefit from its proximity, which they will work first before traveling afield. Bees fly no farther than they must to find food.

All of this assumes, of course, that the area where you plan to locate your bees is not one already saturated by other beekeepers. If this is the case, you have a couple of choices. Find another locale, or try a hive or two and see what happens. Perhaps the neighborhood isn't really saturated. However, if it is, keep in mind that not only will your own bees suffer, but so will those of the other beekeepers that were there first. Consider other options before placing a hive in an area that is obviously already saturated with bees.

Equipment and Supplies

The array of equipment necessary for successful beekeeping is important and you should know and understand it. Standard equipment has evolved over a long period of time. It meets the needs of both the beekeeper and the bees, and for the most part, has proven its worth. Standard equipment has also evolved to provide specific sizes, dimensions, and configurations suitable to the conditions under which bees are kept in this country. None of this should be taken lightly. To repeat a general rule — stay with the tried and true as a novice. Do not experiment. Learn the basics and work with the standards. After 2 or 3 years, when you are beginning to understand bees and beekeeping, experiment if you wish. By then you will have an idea of what you are deviating from, and if things go wrong you will be able to get back on track with minimum trouble.

If you look in any beekeeping catalog you will see a myriad of equipment and supplies that can be confusing, perhaps overwhelming. However, the equipment needed for the first year and even beyond is relatively straightforward. Once you have some experience, you will come to understand the catalog offerings and can begin to pick and choose. A good rule is never to buy any piece of equipment without understanding its use and value. Don't get carried away by someone else's enthusiasm.

In much of this country most beekeepers use two full-depth, ten-frame hive bodies as their basic

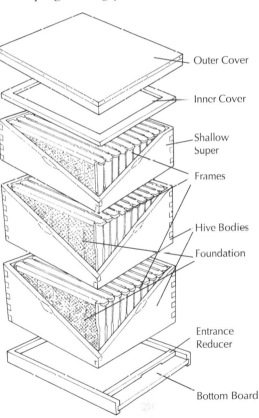

Outer Cover

Inner Cover

Shallow Super

Frames

Hive Bodies

Foundation

Entrance Reducer

Bottom Board

One of the most popular hives configurations — two full-depth, ten-frame hive bodies, with honey supers added during the active beekeeping season.

Basic Beekeeping Equipment

The following list includes everything necessary for the entire first season of keeping bees, assuming a two-story, ten-frame configuration.

Complete Two-Story Hive
- Bottom board with entrance reducer
- Hive bodies (two), full-depth, 9½-inch
- Frames (ten per hive body) full-depth, 9⅛-inch with wedge-style top bar and two-piece or split bottom bar
- Foundation — hook-wired, 8½-inch
- Support pins (four per frame), or frame wire and eyelets
- Inner cover — ventilated
- Outer cover — telescoping

Honey Supers
- Super — shallow, 5¹¹⁄₁₆-inch or mid-depth, 6⅝-inch
- Frames (ten per super) — shallow, 5⅜-inch or mid-depth, 6¼
- Foundation — hook-wired, shallow, 4¾-inch or mid-depth, 5⅝ inch
- Support pins (four per frame), or frame wire and eyelets
  In the long run you will want two or three supers per hive. In the first year you may not need any.

Bees
- 3-pound package of Italian bees with queen

Necessary Equipment
- Hive stand
- Hive tool — 10-inch standard
- Smoker
- Feeder
- Veil and helmet

Optional Equipment
- Gloves
- Coveralls
- Queen excluder

unit, with honey supers (superstructure) added and removed at appropriate times during the active beekeeping season. However, in some areas, a different configuration may be used: one full-depth body, for instance, or one full-depth and one shallow. Although the standard for most of the country is ten-frame hive bodies and supers, and that is what you will see in just about every catalog, narrower and wider boxes do exist. Eight-frame and eleven-frame units are seen occasionally. A beginner will do well to stay away from these less common sizes. Again, in starting out, stay with the standards.

As stated earlier, in some areas of the country different hive configurations may be standard, reflecting the needs or preferences of that particular area. Before acquiring any equipment, find out what the standard is for your area and adjust the list accordingly. But do this because you are convinced that it is a standard and that you will be more or less in step with other beekeepers around you.

The Hive

Now that we have listed all of the equipment you need to get started, let's look at it in some detail, starting with the hive itself. One of the first considerations is the hive's size and shape. Hive bodies and supers can be difficult to handle. When you are lifting an empty box you may not think of this, but when a hive body is full of comb, bees, honey, and pollen it is heavy and unwieldy. A full-depth hive body may weigh as much as 80 pounds, although it is usually less. It is possible to use smaller boxes.

If your size, age, or physical condition limits you, consider that instead of using the standard two deep hive bodies (9½-

The standard two deep (9½-inch) hive bodies (left), and the alternate configuration of three mid-sized (6⅝-inch) hive bodies, give about the same inside volume.

inch), you could use three mid-size boxes (6 ⅝-inch) as an alternative basic hive. The interior volume of the hive is about the same for either. The bees won't care and you will have considerably less weight to lift with each box. Another advantage is that if you then use mid-size boxes as your honey supers, you will be dealing with only one size box throughout the hive, allowing for complete interchangeability of equipment. A disadvantage is that you have three boxes instead of two to manipulate when you are working your hive. This is not something to be overly concerned about if you only have one or two hives. If someday you expect to be into bees in a big way, though, you may want to think about this extra labor. Furthermore, you probably will be out of step with most of the beekeepers around you. This could create very minor problems in that you will not be able to exchange frames or do certain advanced hive manipulations as readily, but these problems are not of sufficient concern to stop you from taking this route if it seems appropriate for you otherwise.

Bottom board. Most of the bottom boards you will come across are the so-called reversible type: they can be used with either side up. One side has a rim of ¾ inch (sometimes ⅞ inch), while the other side is ⅜ inch. The normal position is with the deeper side placed up. The original intention was that the board be reversed (placed shallow side up) for the winter, but few beekeepers do this anymore. Generally, it is not considered necessary. A nonreversible bottom board with a ¾-inch depth is quite acceptable.

Frames and foundation. The list of equipment above is specific. It lists hook-wired foundation and frames with wedge-style top bars and split bottom bars. Although not explicitly stated, foundation made of beeswax and frames made of wood are implied. This reflects a personal prejudice. From experience, I prefer not to use plastic in any part of the hive. Frames and foundation (and some other hive parts) made entirely or partially of plastic are available, but for me they have not proven satisfactory. Many beekeepers do use them, however. Perhaps you will choose to do so.

In my experience, plastic frames break more easily than wood when subjected to pressure from a hive tool, and plastic-based foundation is not as readily rebuilt by the bees if it is damaged or stripped of its wax coating. The bees also seem less willing to work with new plastic-based foundation under certain conditions.

It is very important to acquire foundation and frames that are compatible with each other. For instance, wedge-style top, split bottom frames are specified on the list. Grooved top bar frames are also available, as are grooved bottom bars. These have a place in the scheme of things, but do not work well with hook-wired foundation; they will almost certainly cause the foundation to buckle and bend. If you are going to use any of the alternative frames and foundations, be sure that the various parts are designed to be used with each other.

Pins or wire? Although the recommended foundation is wire reinforced, it is still advisable to fasten it into the frames so that it will not bend or buckle before the bees are able to draw out the foundation by building cells. The traditional way to do this is with frame wire — cross wiring added to the frame itself.

Wiring is not difficult although a couple of tools are necessary and the job can be tedious if many frames are involved. Wire does add some strength and stability to the frames and the resulting comb. However, for most beekeepers, pins will do the job. The important thing is to keep the edges of the foundation straight until the bees have drawn it out and fastened it to the end bars of the frame. If the edges are kept straight, the main body of the foundation tends to follow. I have found that although there are four holes normally drilled in each full-depth end bar, two pins are adequate.

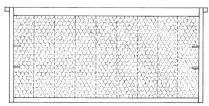

Pins work well for fastening the foundation into the frame. Here, two are placed on each end bar.

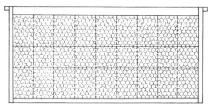

Wiring is an alternate way to attach the foundation to the frame.

Covers. Telescoping outer covers and ventilated inner covers are specified. A telescoping outer cover is one that hangs or telescopes (extends) over the edges of the hive body on which it rests. Its design makes it reasonably weathertight. An alternate cover style, the so-called migratory cover, is used primarily by commercial beekeepers. This style has no overhang on the sides and often, none on the front and back, either. It is less weathertight but allows for hives to be packed more tightly when loading them on a truck. A migratory cover is normally used without an inner cover.

A ventilated inner cover has a top and a bottom and an oblong hole in the center. The hole has several uses, which are discussed in other sections of this book. In normal use, the rim goes up, the flush or flat side goes down. In certain circumstances (wintering, for instance), this position is reversed, as discussed on pages 112–113.

Covers are sometimes made with, or entirely of, plastic or masonite. My experience has been that these sag, so that the inner cover then rests on the frame tops in the center of the hive body. This creates a violation of bee space and the bees will then regularly fasten the covers to the frame tops, which is very disruptive when the hive is opened. I much prefer covers of solid wood. They retain their shape and do not absorb as much moisture as masonite does.

Honey supers. The two sizes of supers listed are the most common and may be considered standard, although other sizes do exist. You should choose one size and stay with it. Avoid any temptation to try one of each. This will just lead to frustration when you find yourself trying to fit the right frames into the wrong box. Speaking from experience, it is surprising how often this happens.

The two boxes differ in size by about 1 inch. The shallow size holds about 30 pounds of honey when full, the mid-depth about 35 pounds. That extra 5 pounds can sometimes feel much heavier. I suggest the shallow super.

Materials and finish. Most wooden hive parts are made of pine although cypress is available from some sources. Both of these woods are quite satisfactory and give years of service. Exposed wood should be painted or otherwise treated to protect it against the elements. Use quality materials for whatever treatment method you choose.

Traditional wisdom says not to paint any inside surface of the hive because the bees don't like it. Take this with a grain of salt. Bees do not object to paint if it is dry and has weathered long enough for odors to dissipate. As a practical matter, I paint the entire bottom board and outer cover. Both of these take abuse from the bees, from the beekeeper, and from the weather. I do not paint any part of the frames or the inner cover. I paint only those surfaces of the hive bodies and supers that are exposed to the weather. Some bee-keepers paint the inner cover and the inner surfaces of hive bodies and supers with no apparent ill effects. (Again, it is important that the paint be thoroughly dry and the odors dissipated before the equipment is given to the bees.)

White is the traditional color for painting hives — it has the advantage of reflecting the sun and helps to keep the hive a little cooler in very hot weather. Dark colors absorb sunlight and can cause the hive to overheat. A disadvantage of white is that the hive stands out starkly against the landscape, making it a more likely target for vandalism or theft. Actually, any light color is satisfactory. I use light green, which helps to camouflage my hives. If your bees are in your own backyard, theft or vandalism probably are not a problem, but if you have outyards, give color and other means of camouflage some thought.

Occasionally the question is raised as to why hardwoods are not used in hive construction. There are two reasons: weight and cost. Hive bodies and supers are already heavy when fully utilized by the bees and most hardwoods are significantly heavier than soft-woods. Hardwoods also usually cost more.

Other Equipment

The hive stand. A hive stand is essential. It is important to keep the wooden bottom board off the ground where it is subject to dampness and rot. However, the conventional hive stand offered and illustrated in equipment catalogs may not be the best option. First, it is made of wood, usually pine, and will rot quickly even if painted. Second, it is narrow, only the width of the hive. As the hive grows taller in the summer months, it becomes a bit less stable and a little more tipsy.

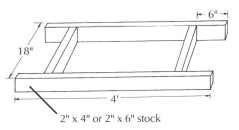

This homemade hive stand, designed with a wide base to accommodate either one or two hives, is more stable than the narrower commercially-available stands.

I prefer a wider-based stand, one that is 18 inches deep and 4 feet wide and made of treated 2-by-4s or 2-by-6s. Treated lumber lasts for years, even in direct contact with the ground. Such a stand can accommodate either one hive, in the middle, or two, one at each end with plenty of room between for you to work. The stand described in this paragraph is not available commercially as far as I know. I make my own.

The hive tool. You could get along without a hive tool, but you don't want to. Its primary use is to slip between hive bodies or supers to pry these hive sections apart. Some beekeepers suggest that you use an old screwdriver or anything else that is handy, but this can severely damage your hives. You would soon find the hive parts chewed up, bees coming out in odd places, and weather and, ultimately, decay getting in. Your hive would tend to deteriorate rap-

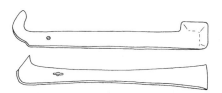

The frame lifter (top) and a standard 10-inch hive tool (bottom) are useful for opening the hive or removing frames without damaging adjacent frames or the hive body.

idly. A conventional 10-inch hive tool will do minimal damage to the hive. This tool also has many uses around the house. Consider buying an extra one for the household toolbox so the hive tool will not be used for other purposes.

A second style of hive tool is also available, the so-called frame lifter. I recommend it highly. It does a wonderful job of removing frames from hives without damaging adjacent frames or the hive body. I generally carry both styles when working hives.

The smoker. A smoker is one of the essential tools of beekeeping. The smoke calms and soothes the bees and makes life at the hive much easier for the beekeeper. You may occasionally encounter a beekeeper who claims never to use one. You can be sure that that individual finds it necessary to be well clad in protective gear.

A 4-by-7-inch size is fine if you have one or two hives. If you have more hives, or plan to have more in the future, the longer-burning 4-by-10-inch size is better. Most smokers today are made of stainless steel. They are more durable and expensive than those made of galvanized metal or tinplate, but are worth the extra cost.

The feeder. Beginner outfits offered through beekeeping

equipment catalogs usually include a Boardman feeder. This style of feeder has a use in beekeeping, but not in a new colony in the spring. One reason is that it is too small; another is that it is used outside the hive where it is exposed to the cold. When feeding is necessary, it should be done in quantity, by the gallon. The Boardman feeder is most often used with a quart jar. It will also accommodate a 5-pound honey jar, but this is less than 2 quarts.

Feeding is usually done in cool weather. The bees will not normally leave their cluster in cool weather and go out to the feeder, so the feed needs to be placed close to them. One method of feeding (described on pages 84 and 88), uses one (or more) gallon glass jars placed on the inner cover. Such jars are used commercially for mayonnaise or pickles and empty ones can often be obtained from restaurants or school cafeterias. They are quite satisfactory as feeders when thoroughly clean. Plastic pails of various sizes, designed for feeder use, are also available through some regular beekeeping equipment sources.

Veil and helmet. The veil is an essential part of your beekeeping attire; other protective gear is less important, depending on your philosophy and approach. You may be able to dispense with a veil occasionally, but most of the time you should wear one. Stings to the face and the head are usually painful and can be dangerous. Most standard veils require a helmet to be worn with them, although some of the newer designs are intended for use without a helmet.

Standard veils are available in two styles, round or square. For me, the square veil is worth the small additional cost. When not in use, it can be removed from the helmet and folded. The round veil must be rolled because it does not fold readily. It can easily become wrinkled and creased so that each time it is used it must be reshaped, a nuisance at the very least, and reshaping causes wear and tear to the veil. If the helmet and the veil will be left as one unit and hung up between uses, none of this matters. The round veil will be fine.

An alternate style, the Alexander veil, is a standard round veil used without a helmet. It has its own built-in soft top. For long-term regular use, this veil is less than satisfactory. It, too, must be reshaped for each use, and without a stiff helmet brim some part of the veil can easily touch the beekeeper's face or head. It is useful as a spare or guest veil, though.

The helmets offered for beekeeper's use are generally suitable, although some may be heavier and a little more cumbersome than others in warm weather. I prefer the nonplastic ones. Actually, any broad-brimmed hat will serve, but it is usually best to dedicate the selected hat to beekeeping use so it will always be available with the veil.

Gloves. It is a good idea to have a pair of gloves for beekeeping use. There are times when the bees become infuriated but you must continue to work with them (for instance, after you have inadvertently dropped a frame or even a hive body while working the hive). You can't just walk away and come back another day. You must finish the job, and gloves are handy. Other than for such a situation, I recommend that you try to do without gloves. You will be more in touch with the bees working bare-handed, less likely to crush and kill them with your fingers as you work, and therefore, less likely to provoke them.

Any kind of gloves that cover you adequately will do, but for regular or long-term use, beekeeper's gloves are advisable. It is nearly impossible for bees to get through them.

Coveralls. Beekeeping catalogs show both coveralls (without a veil attached) and bee suits (with a veil attached with a zipper). Except for the veil, the two are usually identical. Keep in mind that if you buy a bee suit, then you must wear the suit in order to attach the veil. If you buy coveralls and a separate veil, you can wear the veil alone and use its drawstring closure.

Should you wear any kind of suit or coveralls? The same philosophy applies as that mentioned for gloves. The more insulated you are from the bees, the less attuned to them you are likely to be. Try going without. The bees are not out to get you.

Queen excluder. The queen excluder has a place in hive management, but rarely will a first-year beekeeper need one. Buy one if you wish, but put it away for now. Keep in mind that many beekeepers refer to this device as a "honey excluder," and it can, in fact, act as such. One such situation involves a weak nectar flow and supers with foundation. The bees are decidedly reluctant to draw foundation above an excluder and a weak nectar flow compounds the problem. Beginners often deal with both conditions — they commonly use foundation during the first season, and place honey supers on the hive later in the season when nectar flows are slowing down.

Once again, the catalogs are full of equipment, and they can be tempting. Resist the temptation until you are better able to make a first-hand judgment or to ask questions and have the experience to properly evaluate the answers.

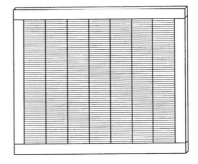

The queen excluder can also be used as a honey excluder, but it is rarely needed in the first year of beekeeping. Resisting the urge to buy too much equipment is one of the challenges for the beginning beekeeper.

Without question, you will eventually want other equipment. Wait — it will still be there in coming seasons.

Acquiring Bees

You have thought about all of the ramifications of becoming a beekeeper, and now it is time to acquire the bees and the equipment. Let's discuss the bees first, because they have a bearing on the initial equipment you'll need, and go over the equipment afterwards. We need to think first about the form these bees will take and then go on to the sources. These sources are not commonly known outside of beekeeping circles, but finding them should fall into place as we proceed.

A beginning beekeeper has at least five options when acquiring bees:

- ✖ A package of bees
- ✖ A nucleus hive
- ✖ An established colony
- ✖ A swarm
- ✖ A wild colony

We will consider all five of these, but I recommend that, as a beginner, you limit your choices to the first two — a package or a

WHAT IS A PACKAGE?

A package is a screened cage containing a specified quantity of bees and a queen. It does not include comb, just bees, which are measured in pounds. Readily available sizes are 2, 3, and 5 pounds. Bees are generally considered to weigh in at about 3,300 per pound, so a 3-pound package contains approximately 10,000 bees.

Package bees usually originate in the southern tier of the United States, with most coming from Florida, Georgia, Alabama, Mississippi, Louisiana, Texas, and California. Package production is limited to these states because of the climate and the length of season. Southern packages are available for shipment in the spring and are sent all over the country via mail or direct trucking. In the more northern states, we could produce packages but they would be available too late in the season for the resulting colonies to be successful.

nucleus. Later, as an experienced beekeeper, you may find yourself using any or all of these methods to add to your holdings.

A Package of Bees

Package bees have long been my personal preference for beginners and they still are, although developments in recent years have made this method a little less desirable. Unfortunately, package bees have been carriers of mites since the mites' incursion into this country. This situation is more under control now, but is always something to be considered.

The advantages to the beginner in using package bees are several. They are readily available, the price is reasonable, but perhaps most important, they are easily handled by a novice. This is because such bees do not comprise a true colony. They are a bunch of bees in a box. They lack a home to defend and they lack organization and sense of purpose. Therefore, they are amongst the least defensive of bees. If you handle them with reasonable care so as not to injure and to crush the bees, you can readily move them into a hive. Usually they accept this hive quickly, but even so, initially they are a small colony, one that is still amenable to handling. Later, as the bee population begins to grow and the bees gain confidence, so does the novice beekeeper. A further advantage to package bees is that, since no comb is included, the possibility of transmitting diseases with them is minimized. Mites are a potential problem, though, and precautions must be taken. (See pages 148–158 for more information on mites.)

Package bees do have disadvantages. One is that it is a slow way to start because it takes time for a package to become established and to begin growing. As a result, there may not be any surplus honey the first season. For most hobbyists, this should not be a major consideration. A second drawback is that a package has absolutely no resources when installed in a new hive and is largely dependent on the beekeeper doing the right thing at the right time. A third disadvantage is that the queen is not an original member of that colony. She has been introduced to the package under somewhat traumatic conditions and has only been with them for a few days. Usually, when she is properly released into the colony after the package has been installed in the hive, she is fully accepted and all goes well. Occasionally, there is a problem and the queen is not accepted. We will consider this further when we discuss installing packages.

A Nucleus Hive

A nucleus hive, or nuc, has long been my second choice as a starter unit. A nuc is usually purchased locally and is drawn from an established hive. It is a small but complete colony with frames, comb, brood, adult bees, and a new, young queen. The frames include, in addition to the brood, stores of honey and pollen. Nucs vary in size, but most common are those with three, four, and five frames. At the time of delivery, nucs are variously housed in a throwaway box or in a returnable box. Sometimes the new beekeeper is expected to go to the source with a single-story hive ready to accept the nuc and carry it home.

The advantages of a nuc are several. First, it is a functioning colony, ready to continue growth and to work in its new home. Second, it has a laying queen who is an accepted member of that colony and there is much less chance of having initial queen problems. Third, because it is a functioning colony with certain initial resources, it stands a good chance of building to the point where it can produce surplus honey in the first season. One further important advantage of a nuc is that, even though it is established and functioning, it is a unit small enough to be minimally defensive. As with a package colony, a new beekeeper can work with a nuc confidently, and can grow as a beekeeper as the colony grows.

A nuc does have at least one disadvantage. Because it is a functioning colony with comb, it may contain disease organisms. But this is not enough of a disadvantage to eliminate nucs from consideration. It is simply something to be aware of and to protect against.

An Established Colony

It would seem natural to get started in beekeeping with an established colony. After all, this is what you are working toward. For a beginner, though, the disadvantages outweigh the advantages. An established colony is one that has been in existence for at least one season, is probably housed in two hive bodies, sometimes more, and contains a relatively large number of bees. Such a colony can be awesome and off-putting to the novice, who may not be able or willing to work with it in the way necessary for the colony to thrive or for the beekeeper to learn.

Secondary disadvantages are also important. Usually, existing hives offered for sale are more than one season old; they are likely to have been around for several. Perhaps the hive has been well tended, is in excellent condition, and seems to be a good value. But reasons for sale may include the owner's declining interest, age, or ill health, any of which could have caused neglect to the hive. The hive or the colony it contains may not be in good condition and the novice does not have the experience to recognize its deficiencies. Possible problems include general debility of the population, disease, or equipment that is defective or poor in quality.

A Swarm

When captured by the beekeeper, a swarm is commonly used to start new colonies. Capturing a swarm is not necessarily difficult. Once in hand, it may be treated much the same as package bees and similar results can be expected. If a swarm is available, this is a perfectly acceptable way to get started. That's the hitch, though — *if* a swarm is available. You cannot count on it. Furthermore, swarm season extends from late spring to early summer. By the time that the swarm season has ended and you realize that you are not going to get one this year, it may be too late to start a package or to find an available nuc.

A further and important disadvantage is that you will probably not know the origin of the swarm and it may carry disease or mites. These are unnecessary problems for a novice to take on. And with the introduction of the Africanized bee to this country, in some areas it may no longer be legal to pick up and hive swarms as new colonies. Your state or province may require that swarms be picked up only by authorized personnel and that they then be destroyed to help prevent the spread of Africanized bees.

A Wild (Feral) Colony

A feral colony (that is, a wild colony living in some location outside the control of a beekeeper) is a final option. It is entirely possible to capture a wild colony and place it in a hive. However, it is usually a lot of work and can be beyond the competence of a beginner. It must also first be available and then accessible. As you will eventually discover, bees have a knack for moving into inaccessible places — in hollow trees high above the ground, in chimneys, and inside the walls of a building are all typical. The days when we could freely cut down bee trees have long passed in most areas, and not many people have the skill or the desire to open up the wall of a building and then neatly close it up again after removing the bees.

Sources of Bees

The sources of feral colonies and of swarms are fairly obvious, and as soon as people realize you are a beekeeper, you will start hearing about them. If you are really interested in picking up these wild colonies, now or in the future, notify your local police and fire departments — householders often call these agencies if they have a perceived or real bee problem, and police and fire officials are very pleased to know of beekeepers who will help. The local exterminator is another source. She or he has many calls to eliminate stinging insects and often does not care to deal with them, so your assistance may be welcome. Remember, though, you may find yourself being called on to deal with wasps and hornets as often as honey bees.

Existing colonies of bees are often listed in newspapers, advertised formally, or learned of by word-of-mouth through beekeeper associations or ads in beekeeping publications. Just keep asking your beekeeper friends and acquaintances — the bees will turn up.

Nucleus hives are usually produced locally, most often by a commercial or sideline beekeeper, but occasionally by a small operator who has extra bees. Ask your beekeeping equipment supplier about where to get nucs, and watch the local beekeeping publications.

Package bees are advertised extensively in beekeeping publications, and local beekeeping equipment dealers often have them as well. Many of these dealers arrange for truckload shipments in the spring, although some of them simply consolidate orders and have them mailed in.

Whatever your choice, don't wait until the last minute. Find your source and place your order at least a month or two in advance. This will usually be in late winter. Last-minute orders are sometimes delayed because of the rush until past the optimum time for starting a colony.

Working with Bees

The idea of simply approaching a beehive, let alone of opening it, brings shivers to many people. Whether this is a learned attitude or something innate is often good for a lively discussion, but either way, many people are tentative about bees. If you have any reservations about actually dealing with bees, probably the best strategy is to just get in there and do it. Find someone to initiate you — an experienced, knowledgeable, patient beekeeper who will open the hive and show you what it is all about. When you start looking at the different kinds of bees, those minute little eggs and the rest of the brood, the stores of honey and pollen, and all of that hustle and bustle, you will quickly forget that you had reservations.

Before you do get into a hive, whether on your own or under guidance, there are some things to think about: when should you work bees, what should you wear, how should you act?

When to Work with Bees

Bees are like people in some respects. (Anthropomorphism rears its head here. Bees are not like people, but saying they are makes some explanations so much easier.) The bees appreciate and enjoy good weather. They are more mellow on nice days. Those are good days to work with the bees. Conversely, a hot and humid day, especially if it is overcast, is not the best one to work with the bees. The reason is not that the bees actually appreciate the weather. It is that the weather controls nectar flows and foraging. On nice days, when the nectar flows, the bees are busy and preoccupied. You become a minor concern in their total scheme of things. On a cloudy day there is usually less of a nectar flow, if any at all. Most of the older bees, the foragers whose venom glands are fully functional, have nothing to do in the hive. You become the big diversion of the day.

Very hot days are best avoided, too. The bees tend to stop foraging as the temperature becomes unusually warm, and they are all hanging around the hive, waiting for you.

What to Wear

This topic has already been discussed to some extent on pages 71–72, but a few more words are in order.

The conventional picture of a beekeeper is of an individual swathed in protection — bee suit, helmet, veil, gloves, and whatever else seems appropriate. Without question, many beekeepers dress this way. However, many others wear little or no protection.

They dress as if for a summer outing, perhaps with a veil, perhaps not. Either approach is acceptable. Individuals wear whatever makes them comfortable, both physically and mentally. Wear whatever makes you comfortable and gives you confidence. To be sensible, though, you should at least wear a veil. The face is vulnerable and usually becomes the first target when bees become upset.

If you elect not to wear a bee suit, keep in mind that the bees have preferences of a sort, and you should know them and dress accordingly. For instance, bees often react adversely to dark colors or to woolly or hairy fabrics. Materials that in any way make you resemble an animal may set them off, especially if they have recently been bothered by a skunk or bear or subjected to any form of harassment. With this in mind, wear light-colored, smooth fabrics. Keep long hair tucked up. Tuck in or otherwise secure your pant cuffs. Bees, especially the young ones, often fall from the frames as you are working the hive. If they fall to the ground around your feet, they are inclined to climb back up rather than fly. You are often the handiest thing for them to climb on and they are as likely to climb inside your pants as outside. They are not trying to get you, they are simply taking the easiest route. Don't let them in and you won't have to get them out. If you tuck pants into socks, white socks are better than dark ones. If you do upset the bees, dark socks become a primary target. Keep in mind that the color red looks black to bees.

Think also of what odors you may be giving off. It is usually best not to wear perfume, hair spray, after-shave lotion, or similar fragrances. The bees are usually not offended by these odors but are attracted to them. They investigate, bump, and bang around your head, make you apprehensive, and often become tangled in clothing or hair. Then the bees panic and you may panic.

Other odors, gasoline for instance, may be offensive to bees. If you have been working on your car, gasoline or other petroleum products often leave a residual odor on or about you to which the bees may react. At the same time, you may find the bees to be very tolerant. There is an old beekeeper's tale that because the alarm odor associated with stinging resembles the odor of bananas, it is dangerous to eat a banana near bees. I tested this one day. I leaned on a hive and watched the activities around the entrance while

HOW TO ACT AROUND BEES

Above all, be calm. Think calmly, move calmly. Movements are important. The nature of bees' vision is such that they see moving objects better than stationary objects. If you remain unmoving, you seem invisible to them. You are not really invisible, but you are not attracting their attention. Conversely, if you create a disturbance and then run or thrash about while swatting at aroused bees, you may attract more attention than if you had stayed still.

peeling and eating a banana. Absolutely no reaction. I laid the empty banana peel down on the entrance board. They ignored it. Some of the bees crawled over it going in and out. Still no reaction. I don't recommend eating a banana near a hive because bees' reactions can be inconsistent, but that was my experience.

A final word about attire. Remove anything shiny or flashy — rings or a wristwatch, for example. Bees are attracted by motion, and these shiny objects glittering in the sun may catch their attention. If you have created any disturbance when you opened the hive, aroused bees may go after the flash of a ring or a bracelet as your hand moves about over the hive.

In sum, neutral fabrics, colors, and odors are best when you are around a hive.

How to Work with Bees

Honey bees do not hear, at least not in any conventional manner. However, they do feel substrate vibration, which means that our movements as we open the hive are transmitted to the bees through their feet. The gentler we are, the less they feel, and the less they are motivated to react.

If your hive is on a broad stand or platform large enough to accommodate you as you work, your movements may be transmitting vibrations for as long as you are there. These vibrations may be minor, but the cumulative effect may be upsetting to the bees. Think of this if you have several hives all on one continuous stand. Each hive may feel the vibrations as you work the others.

Any time that you must thump or bang, do it quickly and be done. For instance, if the outer cover is glued down by propolis, give it one good upward whack to free it, rather than several more moderate raps. Get it over with and the bees may have no reaction at all. They will almost certainly react to persistent banging.

If the bees do get upset while you are working with them, try walking slowly away. Leave the immediate vicinity for a few minutes and give them a chance to settle down. Then go back and try again. Sometimes that is all it takes. Other times, you may find it necessary to close up the hive and continue working with them another time. They have their bad days, too.

If you have difficulty with your bees every time you work them, find an experienced beekeeper to come and watch you work. Maybe he or she can point out something you are doing that is causing your problems. Of course, it could be the bees themselves. Some colonies are just plain mean. (They are not really mean. They have a lower reaction threshold.)

If you have difficulty working with your hive, but only on occasion, stop and analyze the situation. Are you doing anything differently or wearing anything different? Perhaps the weather is oppressive, or perhaps there have been predators or some other form of harassment. It might be your attitude; maybe you don't really want to be there today.

In general, though, treat the hive as a good friend — calmly, respectfully, with kind words and thoughts. Usually, the bees will respond the same way.

Installing Bees

Once you have chosen between a package or a nuc, you must prepare to receive and house your new colony. For both you need basically the same thing: a single-story hive with an appropriate number of frames with foundation and a means of feeding. Presumably, you will be feeding sugar syrup made from table-grade sugar mixed 1 to 1 by volume with water. Prepare at least a gallon now, although you will need 2 or 3 gallons, sometimes more, during the first month or so. Now, with everything ready, let's proceed with installing a package.

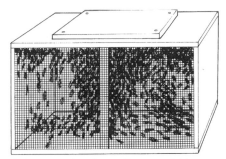

A three-pound package of bees, available commercially, contains about 10,000 bees, including workers, a few drones, and a queen.

Installing Package Bees

Your package should be similar to the one pictured: a screened box containing the requisite number of workers, perhaps a few drones, a queen, and a can of syrup on which the bees have been feeding during their trip. If all is well, the workers will be clustered around the queen and the syrup. The queen is confined in her own small cage near the top of the package, next to the can. You may see some dead bees in the bottom of the package. This is normal. A few bees die during their confinement, but you shouldn't see large numbers of dead bees. If you can still see the wood of the package floor, you may consider the die-off as normal. If there is a thick layer of dead bees beginning to build up, get the package into a hive as quickly as possible.

If you must delay installing the package. You may not be immediately ready to install the package when you first receive it. Perhaps the weather is not cooperative or you must delay for some reason. This is not a problem, but feed the package. The simplest way to do so is to squirt or to spray some syrup onto the screen of the cage. The bees will clean it up quickly. If it gets on the bees, they will simply clean themselves and each other. If you look closely at the screen before you apply feed, you should see many antennae sticking through, and perhaps some feet and tongues. Do not use a paintbrush to apply syrup to the screen as has been suggested by some. By doing so, you might brush off many of those appendages. Instead, tilt the cage to allow the syrup to run slowly down the screen as you spray or dribble it on. Several small applications usually work better than one massive one.

After the package has been fed, put the bees in a quiet, dark, cool (not cold) place. Leave them undisturbed, except for feeding, until you are ready to install them.

Tools and Equipment. Meanwhile, your hive should be set up in its permanent location. Assemble some tools and equipment.

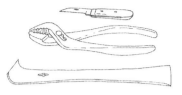

Tools needed for installing a package in your hive include (from top to bottom): a jackknife; a pair of pliers; and a standard hive tool.

The tools you'll need include a standard hive tool for prying the lid from the package, a pair of pliers to remove the syrup can, and a thin nail or a jackknife for puncturing the plug of sugar candy in the queen cage. For equipment you'll just need your regular get-up for working bees: a smoker and your protective gear (a veil at least, but whatever else makes you comfortable). You probably won't need the smoker, but have it lit just in case. Because they have no honey to engorge on, and no hive that they yet recognize to defend, smoking the bees during installation has little positive effect. However, if they become overly upset, or you wish to encourage them to move from one place to another, smoke may help. Keep the smoker handy. Finally, have your feeder ready.

My preference for a feeder is a container 1-gallon or larger, but not more than about 9 inches tall. A 1-gallon glass jar works well. Whatever the container, it should have a tight fitting lid perforated with twelve to fifteen small holes (about ⅛-inch in diameter) through which the bees will take the syrup. Along with the feeder, you will need two small supports, sticks about ⅜ inch thick and 4 to 5 inches long.

Opening the package. Now, put on your veil and any other protective gear. Open the hive, remove both covers, and set them aside. Remove three or four frames near the center of the hive. Lean those frames against the side of the hive away from where you are working. It is very easy to kick them and put a hole through them in the excitement of working with your own bees for the first time, so put them out of the way.

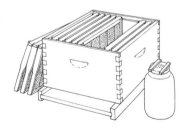

Open the hive, remove three or four frames from the center, and have ready a feeder before opening the package of bees.

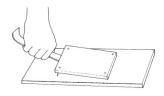

Begin the installation by prying off the package lid, held in place by staples or nails. The feeder can and queen cage inside the lid should keep bees from escaping, but don't be alarmed if a few fly out.

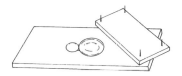

With the lid off the package, the feeding can and queen cage are clearly visible.

Replace the lid, upside down, and bang the package sharply on the ground once to knock the bees to the bottom.

Next, pry the lid from the package. It will be held in place by nails or staples in the four corners. If the feeder can and queen cage are properly positioned, they will keep any bees from escaping as you remove the lid. However, my experience has been that a few bees manage to sneak out as you remove the package cover. Ignore them. They will fly about and buzz, but are not a problem. If many bees are escaping, set the lid back in place temporarily, but turn it over so the staples are not a problem. From this point on in the installation process, however, expect bees in the air. Once you open the package, some of them, perhaps in large numbers, will fly out. In the normal course of events they will not bother you and will eventually find their way into the hive.

With the lid in place, pick up the package and bang it down sharply on the ground or some other solid surface — *once*. Your goal is to knock the bees to the bottom of the package cage so that you can remove the queen cage and the feeder can without releasing hordes of bees. One quick, firm jolt is all that is necessary and does not harm the bees. Do not bounce the package several times; this will just upset them. Now, with the bees in a pile at the bottom, you must work quickly. Some of them recover almost instantly.

Remove the lid and pull out the queen cage and the feeder can. The queen cage will usually be suspended by a metal disk. You may find that the queen cage resists your pulling, so pull harder. Sometimes the bees will have started to build comb inside the package, fastening it to the queen cage and the feeder can. That comb is expendable. Keep pulling until you have the queen cage out. Set it aside, in a protected spot. Still working quickly, use the pliers or your fingers and pull out the feeder can. Set it aside and put the lid back in place, again upside down.

The queen cage is suspended by a metal disk; pull hard on it to remove the cage.

The queen and queen cage. At this point, inspect the queen. You. will usually find her surrounded by a half-dozen bees in the queen cage. These workers are her attendants. They keep her groomed and fed and act as packing material in the cage; they buffer her from bangs and jolts. Sometimes she can be hard to distinguish from the workers, but look closely. She is the one with the longer abdomen. One or more of the bees in the queen cage may have died. This is of no concern as long as the queen is alive.

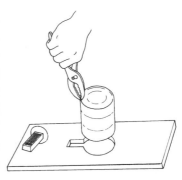

Working quickly with your hands or pliers, pull out the feeder can, and replace the lid.

If you discover at this point that the queen is dead (it happens, although rarely), contact your supplier immediately. A reputable supplier will provide a replacement queen quickly. If the replacement is available locally and you can get another queen right away, delay the installation until she is in hand. If there will be a delay of a couple of days or more, install the package now.

Assuming the queen is well, proceed to prepare the queen cage. One end of the queen cage will contain semisolid white candy. This is their food supply for the trip and also provides a means for a delayed

The queen cage, containing the queen and a half-dozen worker bees, should be examined closely to be sure that the queen is alive.

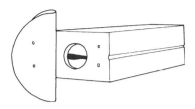

Use a thin nail or the small blade of a jackknife to push a hole through the exposed candy in the end of the queen cage.

release of the queen. It is not desirable to release the queen immediately when the package is first hived. She could be lost or killed in the confusion.

Bend back the metal disk to expose the small hole in the end of the queen cage. Do not totally remove the disk. Candy should be visible through the hole.

However, be careful. Sometimes, if the queen and workers have been confined for a long time, the candy may be partially or entirely eaten and the bees could escape. In that case, you need to replace the candy with something else. Queen cage candy is a special formulation and not readily available, but a piece of marshmallow will do the job. Stuff some in.

Most often you will find that the original candy is intact. It's a big job for the bees to eat it all, so give them a hand. Use a small nail or piece of wire perhaps ¹⁄₁₆-inch thick, or use the small blade of a jackknife, to push a small hole through the candy. The reason for this is not to open it entirely, but to give the bees a place to get their teeth in.

Assuming that all is well with the queen, hang the queen cage in the hive between two of the frames. Use the same metal disc that held the queen cage in place in the package. Be sure that the candy end is up, so that the bees will exit the cage at the top end. This

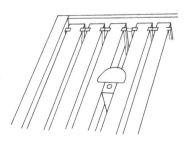

Hang the queen cage between two frames near, but not directly in, the center of the hive. Keep the candy end up so the bees can exit when ready.

way, if any of the attendants die in the queen cage before release, they will fall to the bottom and not block the candy, thereby preventing release. You should place the queen cage near, but not exactly in, the center of the hive.

Presumably, you are installing your package in a hive with foundation in the frames rather than in a hive with drawn comb. However, if you do have drawn comb, you may find it necessary to cut away a small amount of the comb where the queen cage is placed so that the bees of the colony can communicate with the queen prior to her release. Do not allow the screen of the queen cage to be jammed into the comb.

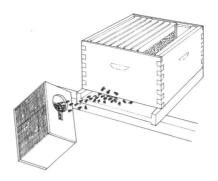

Dumping the bees. With the queen in place, it's time to dump in the bees. Pick up the package container and once again give it a good solid thump on the ground or other surface. Then place it over the hive and pour the bees out into the space in the hive where you earlier removed several frames. Bees will immediately start flying about. Ignore them, but work quickly. You will not be able to shake all of the bees from the package; your goal is to get the majority out. With that accomplished, place the package container at the front of the hive so that the remaining bees can find

After shaking the package container directly over the hive, prop it against the hive entrance so that any remaining bees find their way in .

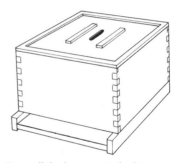

Once all the bees are in the hive, carefully replace the removed frames and the inner cover.

their way out and to the hive entrance. It may take them some time to vacate the package but eventually they will.

You now have a heap of bees in the hive. Carefully replace the previously removed frames, but don't force them down or you may crush bees. Just put the frames in place and watch them slowly settle as the bees get out from under. With all the frames in place, put the inner cover on.

Placing the feeder. With the bees in the hive and all the frames inside, it is then time to place the feeder. Invert it over the hole in the inner cover, on the two sticks, which allows the bees to come up under the lid of the jar where they will suck the syrup from the holes. The syrup should not run freely out of the feeder. To check

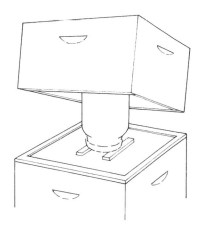

With the feeder jar in place, an empty hive body and outer cover complete the installation.

this, invert the feeder while holding it away from the hive so that spillage will go on the ground. A little syrup will run out initially until an equilibrium is reached, then atmospheric pressure will keep the syrup in the feeder. If the syrup keeps running out when you first invert, either the lid is not tight or the holes are too large. Either way, air is leaking in. Tighten the lid and try again. A leaky feeder can cause several problems: wet bees (as the syrup runs through the hive), ants, robber bees, and a loss of feed, to name a few.

With the feeder now properly installed, set the second deep hive body in place. If you find that your feeder is too tall (it does not allow the outer cover to sit firmly in place), you can use two empty shallow or mid-depth supers instead of a hive body. This will give 2 or 3 inches of additional height for the feeder. Set the outer cover on and installation is complete.

Installing A Nucleus Hive

When you buy a nucleus hive (a nuc), you may go to the source with your single-story hive and have the nuc installed there by the seller. Installation is sometimes done on the spot; other times you drop off your hive and pick it up a day or two later with the nuc installed. After installation, the hive is stapled and screened for travel and off you go. When you arrive home, place the hive in its permanent location, add a feeder, and you're done.

If your nuc is delivered to you in a temporary container, your first job is to move it to its permanent hive. The preparation is much the same as that for a package. Have the hive ready and in place, with its bottom board, hive body, and covers. Have enough frames handy so that, counting the frames that come with the nuc, the total is ten.

Use smoke and open the nuc, just as if you were opening a

regular hive. Bees will immediately fly out. This is normal, so ignore them. Transfer the frames. Maintain the same frame position you find in the nuc. In other words, don't turn any of them end for end or switch their positions. Place them more or less in the center of the hive, then put in the remaining frames to bring it to ten. (This description assumes that the queen for this nuc was installed some days earlier and that she was free on the frames as you made the transfer. If for any reason this is not so and the queen has been delivered to you in a separate queen cage, after transferring the frames, install the queen just as you would with a package.) Finally, put the inner cover on, place the feeder in the manner described for a package, and put on the outer cover.

Initial Management

The first month of existence for your new colony is a critical period. Your actions or inactions during this time can make the difference in the colony's survival over the long run. You must check on it regularly and be alert to any developing problems.

A 3-pound package contains approximately 10,000 bees, which is an adequate number to start a new colony. With the beekeeper's understanding and assistance, the package should be successful. However, packages can fail. The beekeeper must monitor its progress and provide help when necessary.

When a package is made up in a southern beeyard, it receives bees from usually one, sometimes two, colonies in that yard. These bees are taken from their parent colony, separated from their queen, and shaken into the package cage. A new, young queen and a can of syrup are added, and they begin their trek. The bees are in this package for at least 2 or 3 days, but usually longer. It is obviously an abnormal situation for the bees.

The new queen is a complete stranger to these bees. She is young, recently mated, and was taken from her parent colony at least as abruptly as were the bees of the package. As described earlier, the queen travels in her own small cage within the package, separated from the mass of bees by screening. Although they are separated, the queen's odor circulates, her presence is known, and the bees all soon become acquainted. The bees forget their parent queen within a day or so and this new queen becomes theirs.

The trip north is traumatic. The bees travel by various means — by truck, by plane, perhaps even by train. They are constantly jostled and jounced and often are exposed to extremes of weather. For all of that, provided they receive reasonable protection from the weather and don't run out of food, the bees arrive in remarkable condition.

Then we put them in a hive, most often early in the season when there is little forage available. Usually, with new beekeepers, the bees are hived on beeswax foundation. They must secrete additional wax and build comb on this foundation. Wax secretion depends on the continued stimulation of wax glands, which results from a steady supply of nectar (or syrup) coming into the hive.

Several things are working against the bees at this point. As was mentioned, there may not yet be a nectar flow in progress. Even if there is, the weather may still be cool, shortening forage time. Next, they have little or no comb in which to store honey and pollen or in which to raise brood. Also, the queen has been confined. She must be released and begin laying, a process that can take at least 2 or 3 days. Even then, her egg laying rate will start off slowly, building up

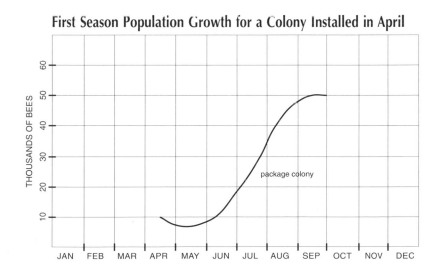

First Season Population Growth for a Colony Installed in April

package colony

as she and the colony become established. It takes weeks for her to reach her full egg-laying potential.

To further complicate the situation, bees are dying. The average life span for a worker during the active season is no more than 4 to 5 weeks. The bees of the original package can be considered just an average group of bees, some old, some young. Bees start to die off from natural causes almost immediately. No new bees are available because it takes 3 weeks for an egg to hatch, to develop through the pupal and larval stages, and to emerge as an adult bee. The result of all this is a slowly declining population during the first month as a new colony.

It is up to the beekeeper to recognize these problems and to help the package colony along. The first concern is the queen. When the package was installed, the queen was not released directly into the main body of bees. This was in recognition of possible danger to the queen. Installing the package is traumatic for the bees. It involves thumping, banging, smoke, and confusion, on top of the stress of the packaging process and the trip. In such situations, the bees sometimes seem to blame the queen, and as a result, kill her if she is released directly and immediately. Using a delayed queen release (as described on pages 86-87) gives the colony time to settle in and become reacquainted with the queen in her cage and in the calmness of its new home. When the candy has been eaten away and the queen is released a day or two later, she is usually welcomed eagerly. Our first concern is that the queen's release happen on schedule.

The First Inspection

During the first few days of its existence, don't unduly disturb the new colony. Too much attention could result in the bees absconding or the queen being killed. The first couple of inspections should be relatively short and specific. On the first inspection, your purpose is to check on the queen. Has she been released? We have been proceeding on the assumption that the candy would be eaten away and the queen released by the third day after package installation. Checking that this has indeed happened is relatively simple and quick, although it does require that you open the hive.

This initial inspection should take place on about the third day after installation. First, light your smoker. You probably won't need it, but have it ready (it is good practice to light it, at any rate). Smokers are simple things, but are often ornery and can go out with no warning. Always light the smoker before you do anything else. When it seems to be burning well, set it aside. Next, don your veil and any other protective gear.

Proceed to the hive. Stop and look at the activity at the hive entrance. Is there a deliberate coming and going of bees, which is evidence of foraging in progress? Are some of them carrying pollen pellets on their hind legs? Do the bees around the entrance look calm, just following the daily routine? Now, take a moment to review exactly what you will do when you open the hive. Make this a meaningful visit. Don't open the hive and then blunder around in there wondering what to do next.

Check the smoker. Is it still burning? If it goes out, it usually happens during this first interval. If the smoker has gone out, you now have the opportunity to relight it before the hive is open.

Once you have established that your smoker is working well, puff it once or twice into the hive entrance. Then, lift the outer cover and puff once or twice under there. Close the cover and step back for a few moments. Allow a little time for the smoke to circulate and for the bees to realize they have been smoked. Remove the outer cover, the feeder, and the inner cover. Another puff of smoke may be in order if the bees seem agitated, but keep in mind that smoking can be easily overdone.

Carefully remove the queen cage. Note the condition of the candy and look for bees still inside the cage. Ideally, the candy is completely gone and the queen is out. There may be one or two bees on or in the cage: it has an attraction to the bees for at least a few hours after the queen leaves. Assuming the queen is out, set the cage aside; you have no further need for it now. But, save it. You may use it in the future for other purposes.

Examine where the queen cage was between the frames. Have the bees started to build some wild comb in that space? This often happens because of the excess space needed to fit in the queen cage — a violation of bee space. Remove any such comb and reposition the frames so that bee space is once again reestablished. If you don't remove that comb now, the bees will continue to build on it, causing significant problems. You may have to sacrifice a small amount of brood now if the queen has laid eggs in this wild comb, but take the loss. Otherwise, you may later find that the two adjacent frames have become joined by bridge comb and have to be handled as one. Your only recourse then may be to replace the entire comb of both frames with new foundation, so don't delay in removing wild comb.

You have now accomplished the purpose of this first hive inspection: the queen has been released and for the moment you presume that all is well. Do take a quick look at the brood area, though. There is no need to remove frames, but if you look down between the frames you should be able to see a concentration of bees and activity in two or three frames, perhaps even in four. This activity usually centers around the area where the queen cage was located. Looking down, you should be able to see evidence of comb being drawn. Note how much area is covered. File it in your mind for future reference. You will want to see on your subsequent hive inspections that this area is expanding.

Now, put the inner cover back in place. Refill the feeder as necessary and replace it. Any comb you had to cut from where the queen cage was may be placed with the feeder. The bees will clean out any nectar or honey that may be in it, taking it below. You can later dispose of the clean wax. Put the hive body and outer cover in place. You're done for today. Plan to return in another 3 or 4 days.

The Second Inspection

The goal of the second inspection is to determine if the queen is laying. You should also check that there continues to be progress in the collection of nectar and pollen and in foundation being drawn.

As before, do an exterior inspection, then smoke and open the hive. Remove the feeder and get down to the frames. With the inner cover off, make a visual check before you remove frames and disturb the organization of the brood nest. Does it look at least as active and as large as before? It is probably early to see any significant increase in development over your last visit, but is the working area at least as big?

Remove an outer frame. Each time you remove frames you cannot avoid squeezing a few bees, perhaps killing them. If you start with an outer frame, you are working in an area of the hive where there are fewer bees. You will do less damage.

With the first frame out, look it over carefully. It probably won't look much different than it did when you first put it in. The bees haven't gotten to the outside yet. Set the frame down, leaning it against the outside of the hive, but away from you so you don't kick it. Try to put it in the shade. By removing that frame, you have created some working space in the hive. Remove two frames to make even more space. It will be easier to remove other frames after the first frame is out because you can separate them from each other and not roll and crush bees as you get into the center of activity. Use your hive tool to separate and slide the frames apart. Pull one out from the center of activity. A frame adjacent to where the queen cage was located earlier is a good choice. That is where the colony probably first started to draw comb and where the queen would first start laying. Look carefully in the cells of this comb for eggs or larvae. Eggs and young larvae can be difficult to see. It helps to stand with the sun coming over your shoulder so that it shines into the bottoms of the cells as you hold the frame angled in front of you.

If you see either eggs or larvae, or both, all is well — the queen is working. She may even be on the frame you are examining. Don't spend a lot of time searching for her, though. You have seen the evidence of her presence and that is what's most important. Reassemble the hive, putting everything back as you found it. Do not change the position of any of the frames. Although it may not be immediately obvious to you, the bees have everything organized efficiently for their needs. Changing that organization will only set them back because they may stop productive work to try and reestablish things as they were.

Eggs (top cells) and young larvae (bottom two rows of cells) in the frames are a good sign that the queen is working and the colony is growing.

If you cannot find any brood, and if the queen was released on time as determined by your previous inspection, perhaps all is not well. It is not yet time to panic, however. Give some thought to where you can acquire another queen — perhaps from the same source where you got the package or from a local supplier. Don't get one yet, though. Wait another day or two, then inspect the hive again, looking for brood as before. Sometimes queens are a little slow in getting started. If on this next inspection you still find no eggs or larvae, assume that your colony is queenless and acquire another one as soon as possible. Install her as before and then proceed.

Subsequent Inspections

Assuming that all went well and the queen is laying, it is time to begin monitoring progress in the hive. Your short-term goal is to have the colony continue to grow from week to week; your long-term goal is to have the colony ready for its first winter. To be sure that this is happening, plan to inspect the colony weekly. This is more frequent than is necessary for an established colony, but yours is far from established and you want to catch any potential problems early. External inspections can, and perhaps should, be more frequent.

Your weekly internal inspections should be relatively brief. There is no need to remove every frame or search for the queen. A visual sweep across the tops of the frames should tell you if the brood nest and attendant activity shows continual growth. Removing and inspecting one or two frames will allow you to inspect the brood pattern and assure yourself that the queen is continuing to lay in good order. Occasionally, however, make a more detailed inspection to become better acquainted with life in the hive.

Feeding

As a part of your inspection procedure from the start, it is important that you check the feeder regularly and never allow it to run dry. The bees do not know the source of the syrup; they do not distinguish between it and a nectar flow. They are only aware that food is arriving in the hive and they respond to it by secreting wax, drawing comb, raising brood, and all of the other activities in a busy colony that depend on an adequate supply of food. If this supply is shut off abruptly, as when the feeder goes dry, the bees may reduce the level of brood rearing or in the extreme, remove some of the brood. At this critical time in the life of your new colony, brood development is paramount.

Every day or two, carefully lift the outer cover and check the level of feed. Refill the container before it becomes empty. If you know you will be unable to check the feed for several days, you could put an extra container in place, feeding 2 gallons at once instead of 1 gallon.

You can usually check and even refill the feeder without using smoke or disturbing the bees. Lift the cover, observe, take out the feeder (if necessary), blow or brush any bees from the jar lid, and replace the cover. Wearing a veil while doing this is probably unnecessary, but is always prudent. You never can tell when you may bang or thump something inadvertently.

Expanding the Hive

After about a month, during which feeding has probably continued, it is time to put the second hive body in its proper place. The timing for this is flexible and depends on how quickly the bees have drawn out and occupied the frames of the first hive body. As a rule of thumb, watch for the colony to be using seven to eight frames in that first hive body within 4 weeks. This does not mean that these frames must be fully occupied or even fully drawn, only that they are receiving a reasonable level of attention and use. At this point, remove both covers and the feeder. Place the second hive body with its ten frames of foundation directly on the first body, and replace the covers. The bees will soon occupy both hive bodies. In fact, you may find that they have expanded into the second story without ever finishing the comb of the outer frames in the lower body. This is something you must watch for and guard against. It is imperative that the bees finish drawing comb in all twenty frames before this first season ends. They will need all of that comb as storage space for the upcoming winter.

If, after a period, you see that the outer frames in both hive bodies are not being properly drawn or used, it is time for you to interfere by shifting some of the frames. Place one or two of those underused frames closer to the center of activity, but don't make any massive disruptions to the brood nest. Swapping the outermost frame with the one next to it may be sufficient for now; further shifting can be done in subsequent weeks as necessary.

Meanwhile, you must continue to think about feeding. You may determine that it is no longer necessary. If the bees have stopped taking the syrup or if consumption has been low, you can stop.

If they are still taking the syrup at a good rate, though, continued feeding is in order. If you have been using the second hive body to hold the feeder jar and do not have a spare body, you must make an alternate arrangement. You have at least two possibilities. A couple of empty shallow or mid-sized supers can replace the empty hive body you have been using, or you can use a Boardman feeder.

Earlier, I expressed reservations about the Boardman feeder. The two concerns were accessibility in cool weather and size. Neither is as important at this point. The weather is warmer and the colony is less dependent on the feed. One word of caution, however. Your colony is still relatively weak and somewhat susceptible to robbery by stronger colonies. A Boardman feeder is often a catalyst in such a situation because it is located outside the hive where it

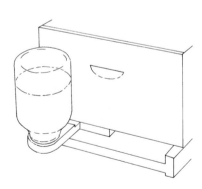

may drip and attract stray bees. To help control this situation, watch for spillage and leaks and buffer the entrance. A section of entrance reducer can be fitted so that the hive entrance, in addition to being small, is not next to the feeder. Any potential robbers are thus forced to actually enter the hive to get at the syrup, where the bees of the colony are better able to defend their space.

If a Boardman feeder is installed, it should be buffered to protect against robber bees. The buffer (to the right of the feeder) narrows the hive entrance and forces potential robbers to actually enter the hive.

At some point, usually by midsummer, you should be able to relax a little and realize that all has gone well. Your colony is established. It is time to move on to the next phase, summer management.

CHAPTER 4

ONGOING MANAGEMENT

The First Summer

O nce your colony has settled down, you see that new young bees are emerging, and the population is on an upswing, you can relax a little. Only a little, though. You must be *sure* that the colony's population is increasing and that all is going well. Don't just hope or assume that all is well. Watch carefully and as you observe and learn, be prepared to support the colony if it needs help. To decide this, look at the growth curve you can expect as the season progresses (as diagramed on the page 91), which will give you some basis for comparison.

During the first month or so the population of your colony was falling. This reflected the natural die-off of adult bees from the original package and the 3 to 4 weeks' delay before the emergence of new bees. At a point during the second month the curve starts an upturn, and this growth should continue until late summer. One of your primary responsibilities from the outset is to monitor the colony's progress and be sure that it does happen, as discussed in the previous chapter. This progress must take place if the colony is to be ready for winter. If it is unduly delayed, the likelihood of the colony surviving the first winter is greatly reduced.

Weekly Visits

Plan to visit your colony at least weekly during its inaugural summer. When you first arrive at the hive, make an external observation. Does everything appear normal — foragers coming and going, the bees calm and preoccupied with their work, no signs of agitation around the entrance? Look for bees that alight heavily, thump down, and walk on in; presumably they are carrying a load of nectar. Watch for other bees arriving with pollen on their hind legs. During the active part of the day an equal number of bees should be leaving the hive.

First Season Population Growth for a Colony Installed in April

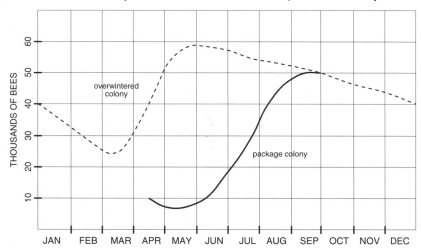

We have already talked about the population rising, and you should check on that once the hive is open, but first, is the weight of the hive increasing? With experience you can get some sense of this. Each time you visit the hive, whether you are planning to open it or not, reach down and lift the rear a couple of inches, enough to get a sense of the relative weight. The first few times you do this may not tell you much, but over time you will begin to discern the differences. Tied to the time of year and the time within the season, relative hive weight should be somewhat predictable. If there seems to be a deviation from what you expect, look more carefully in the hive for positive or negative indications.

An exceptionally heavy hive all of a sudden suggests the need for a super; if it seems a bit light, perhaps you need to feed. As time goes by and you gain experience, you may derive even more information from this simple lifting.

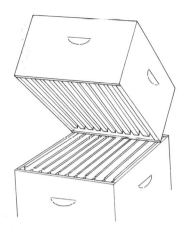

Once you have made your external inspection, open the hive. With the covers off, peer down between the frame top bars. Make this a habit. You may not learn much the first few times you do this, but eventually what you see will have meaning so that you do not need to pull frames every

By opening the hive, and tilting the upper hive body forward, you can observe the bees' activities from the top or bottom. This inspection method eliminates the need to pull frames every time, which disturbs the bees and slows production.

time. Doing so does disrupt activities and can set back production. You can just view the activities from the top or from underneath. It is surprising how much you can determine by peeking down between the top bars of the frames or by looking up under a hive body that has been tilted forward. On these early visits, though, and often later, you will want to pull frames to continue your inspection.

Your main concerns are that the colony continue to grow, that the brood nest expand, and that frames be drawn. You can reasonably infer that if the brood nest is expanding and you can see that the bees are occupying and using more frame space at each visit, then population is expanding. If you can see progress since your last visit, a visual inspection may be all that you need to do this time. On some of these visits, though, you will want to pull some frames, examine the brood, and check on progress in general. Is the queen laying a good solid pattern? Are there several frames of brood? Is there pollen? Is there any sign of disease?

As you become more confident and experienced, spend a little more time studying the contents of the frames and the activities on them. Identify eggs, the larvae in their different stages of development, and the sealed cells containing pupae. Be sure you know the difference between capped brood and capped honey. You should be able to recognize each by both their appearance and locations. Learn the appearance of pollen in the cells. Pollen can be deceptive because of its varied colors — not just the yellows and oranges that we see so often, but the shades of gray, brown, red, green, blue, and white that sometimes appear as well.

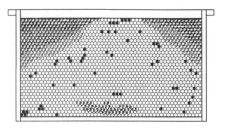

A frame with typical brood pattern shows a solid pattern of capped brood (note the few empty spots).

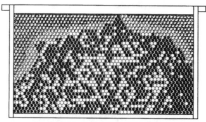

A frame with a poor brood pattern shows an irregular egg laying pattern and many uncapped areas.

Watch the actions of the bees as they move over the frame. Try to identify what individual bees are doing. Pick one and follow it for a while. (Caution: Do not expose a brood frame to excess wind or direct sunlight for prolonged periods. Larvae are tough little creatures, but they can dry out from the wind and they may be harmed by direct exposure to the sun of more than a minute or two.) As much as possible, hold the frames over the hive while you examine them. Young bees that have not yet learned to fly well will land in the hive rather than on the ground if they fall from the frame.

Adding a Honey Super

Presumably, at this point progress is satisfactory. The hive is reasonably well established, population is growing, and the bees are occupying the best part of two hive bodies. They have been drawing out the foundation steadily and it is time to decide if you should put on a honey super. Several things go into this decision. Are you still feeding? How much of the original foundation in the two hive bodies is drawn? Is all of the resulting comb being used?

How far into the season is it? What is the nature of the nectar flows in your area? We will take each of these considerations in order, but the answers are somewhat interdependent.

Most new colonies will stop taking feed as the primary nectar flow comes in and the hive population builds up. This is usually more or less simultaneous with putting the second hive body in place, so you will probably have stopped feeding before the question of a honey super arises. If you are still feeding, it is time to stop. You should not feed while honey supers are on the hive. If you believe that feeding is still necessary and that the hive needs help, then you should not put a super on.

It is best to stay a little ahead of the bees when supering. In placing the second hive body, you looked for activity in seven or eight frames of the first body. You did not wait for all ten frames to be drawn and used before giving them more space. The same holds true for supering. When you see activity in seven or eight frames of the second hive body, consider adding a super. Again, you may wish to manipulate frames in the hive bodies to ensure that the bees give all twenty their full attention; they must be drawn before the end of the nectar flow. If you place a super while there are undrawn frames on the outer edges of the hive bodies, the bees may move up into the super and ignore those frames. They have a tendency to move upward rather than outward as they expand within the hive.

Consider the timing of the nectar flow or flows in your area. Where in the season are you relative to this flow? Nectar flows differ from area to area, of course. Some places have a strong flow in late spring and early summer, ending perhaps in July with nothing significant after that. Other areas have a similar flow, followed by a period of dearth during midsummer, followed by a so-called fall flow. Yet other areas have a more or less continuous flow over the entire active season. Find out from the experienced beekeepers in your area just what nectar flow pattern to expect and the approximate dates. In my area, for instance, there is a late spring/early summer flow and a fall flow, but little or no nectar is available from mid-July until mid-August.

As a new beekeeper, you are probably dealing with foundation in your supers rather than drawn comb. In order for that comb to be drawn, there must be a nectar flow in progress. The incoming nectar stimulates wax production in the hive. Simply stated: no nectar, no wax. Know the timing of the nectar flow for your area and factor it into your decision. If you put foundation on a hive when no nectar is coming in, expect it to be largely ignored, or chewed up and removed to be used for capping cells in other areas of the hive. Also be aware that even though a nectar flow may be in progress, the colony is less inclined to draw comb in the late season when the bees are no longer in an expansion mode.

One further consideration. If a nectar flow is in progress, but you have reason to believe it will soon be over, be very hesitant to place a super. You do not want them to work up the foundation in the super on that last bit of nectar flow while foundation remains undrawn in the hive bodies.

Supering

We have been talking about supering, without ever defining it. In its simplest terms, supering is the placement of the superstructure (supers) on a hive so the bees will have specific space in which to store surplus honey, the honey the beekeeper ultimately will take. Supering can be very simple — when the bees need more storage space, give them a super — or it can be complex. What we are discussing here is the progressive placement of the second super, and the third, and perhaps a fourth, or more. It is unlikely that you will need more than one or two supers in your first year with a new colony, but there are always exceptions.

All kinds of schemes, methods, and systems have been devised to help the beekeeper know where on the hive to place the next super relative to those already in position. A particular method or system of supering should not be an issue during your first summer — keep it simple. Put on one super. If you need another, put it on top. The bees will find it, and if nectar is coming in, they will fill it.

With experience, perhaps you will decide that the method of supering is never to become an issue. Some of the systems used or suggested are far more complex or labor intensive than the results justify. In the long run you may come to believe that keeping it simple is best. When another super is needed, put it on top. However, a decision on this can wait until next season. In the meantime, you are gaining experience and presumably reading, talking, and asking questions so you will have a better foundation on which to base your decision about supering.

Fall Management

The term "fall management" may be a misnomer. We really mean preparation for winter — a process that starts in the spring and continues until the season is over. There are specific fall management tasks, but even some of these may be carried out in August, hardly autumn on the calendar. Then again, in some of the more northern regions, perhaps it is fall in August. For all of this to make sense, we need to jump ahead a bit, take a look at winter, and then come back and make the preparations.

Winter is hard on a colony of bees. They are confined to the hive for prolonged periods, at least weeks, if not for months, but activity still continues. The bees do not hibernate, but carry on with much of their normal routine throughout the winter. No foraging takes place, of course, and the queen does take a vacation from egg laying for several weeks in early winter. Then, during some of the coldest weather, she begins to lay again and brood is raised. If there is a warm day, the bees will fly out to defecate, to drag accumulated debris and dead bees from the hive, perhaps just to get a little exercise and check on the progress of spring.

Bees are cold blooded. An individual bee is immobilized when its body temperature reaches about 45°F (7°C). As the temperature begins to drop through the sixties, the bees begin to cluster for mutual warmth. Initially the cluster is large and loose, but it tightens as the temperature continues to fall. Within the center of the cluster the temperature may be 94°F (34°C) or higher. Here life goes on, no matter what the outside temperature. On the surface of the cluster the temperature can be as low as 45°F. At and near this surface the bees are inactive. As winter moves along the bees remain in the cluster, which expands and contracts with temperature variations. During the warmer weather of early spring the cluster becomes looser, expanding to allow for greater activity within its confines. In other words, as spring progresses, the cluster enlarges and more brood rearing can take place within its protection. Thus, population and the scope of the colony's activities can grow.

To help them prepare for and get through this difficult time, we should give thought to five factors: population of the colony; health of the colony; the queen; food stores; and protection from the elements.

Population

Referring back to the chart on page 102, we know that colony population is going to fall off during the winter. This is from the natural die-off of aging bees, coupled with the reduced rate of brood rearing. However, it is not an abrupt drop — even in the depths of winter there should always be a substantial number of bees present. There are two reasons to think about this population number and be concerned that it is maintained. The first is heat (keeping warm during the winter), and the second is spring growth.

Individual bees cannot keep themselves warm when temperatures fall because they are cold-blooded. However, as a group they can generate and conserve heat within the cluster. This is done by the mass of individual bees as they shiver their flight muscles, fueled by stores of honey. It is this concerted action that keeps the temperatures in the cluster so high during brood rearing, even in the coldest weather.

For the cluster to maintain this temperature, however, there is a minimum number of bees that must be present. If the number of bees fall below that minimum, the cluster cannot keep itself warm — the ratio of volume to surface area becomes too low, heat dissipates from the surface of the cluster faster than it can be generated from within, and the cluster may then freeze. Short of freezing, in extreme cold the cluster may contract so much that some of the brood can no longer be covered adequately. The exposed brood may then chill and die.

Another potential problem in very cold weather is that the cluster may contract to a point that the bees are no longer in contact with their honey stores. They may eat all that is within reach and then become immobilized by the cold, unable to move to nearby honey even though it may be only an inch or two away. A larger cluster of bees that generates more heat can alleviate this problem.

One of the key activities of the new beekeeper over the first summer will be to monitor the population growth of the colony to ensure that the colony does progress. Presumably, the colony has drawn out and is occupying all twenty frames of the two hive bodies. If population seems low, there are at least two possible remedies. One is to requeen. A new queen, introduced early enough, will lay more vigorously than the old one, giving a population burst of new bees in the late season. These young bees not only bolster the population, but create a younger overall population, making the probability of a successful winter better.

Another possibility for increasing colony strength is to give the colony a frame or two of capped brood from a stronger hive. This has the same end result as requeening, but is more immediate since this brood will emerge in a matter of days.

Colony Health

New equipment, vigorous young bees, and ample attention and care during this first season should bring you to the beginning of fall with a healthy colony. However, despite the best care and attention, bees are susceptible to a variety of ills. Two diseases and two mites are of specific concern as we approach winter: American foulbrood, nosema disease, tracheal mites, and Varroa mites. Depending on the absence or presence of any of these in your colony or in your neighborhood, it may be necessary to medicate against any or all of them. Even if none of these are actually present in your hive, you may still wish to take precautions against them. The specifics of diseases and mites, and of medication against them, are discussed in Chapter 6.

The Queen

It is worthwhile to evaluate the queen before each winter and to give some thought to requeening. This is a practice that is carried out annually or biannually by many beekeepers. Although requeening can be done at almost any time during the active season, it is usually done in spring or fall. The term "fall requeening," though, is misleading because it should be done in late summer. There are at least two reasons for this. First, requeening at any time is somewhat chancy. The new queen may not be accepted and it may be necessary to try again. It usually takes a little time to requeen and then to discover if that queen has been accepted. If requeening is done in the fall (September or later), there may not be time to try again if anything does go wrong. You stand the danger of entering winter with no queen and no resources within the colony from which to raise one.

A second reason for so-called fall requeening is the new young brood that results. It is best to have that new queen introduced and laying well before the nectar flow ends for the year. More or less concurrently with the end of the nectar flow, the queen will reduce her egg laying.

Aside from the effects on population, this is the time to evaluate the queen's performance. Perhaps population is good, but for other reasons you have been wondering about the queen. A poor brood pattern, for instance, or perhaps the general disposition of the bees is poor. This is a good time to requeen, correcting the problem you have noted over the previous weeks and, as an added benefit, boosting the population a little more. The resulting young bees are always welcome.

An additional benefit to fall requeening is that a new queen at this time reduces the possibility of swarming in the coming season. It is accepted that a colony with a new queen is unlikely to swarm. By spring, this queen is not brand new, but she is still young enough to have some benefit in swarm prevention.

Food Stores

As summer comes to an end and fall moves in, the nectar flow ends. In fact, in some areas, it ended much earlier. Before the serious cold weather sets in and stops flight activity, it is necessary to evaluate the amount of stores in the hive. There is a requisite amount necessary to get a colony through the winter. This amount varies throughout the country, tied to the specific climate of a given region. Obviously, shorter warmer winters require less food storage than do longer cooler ones. You can determine the recommended amount for your area by asking other local beekeepers.

In much of the country, 60 to 70 pounds of honey is required. This amount is easily stored in a hive made up of two deep hive bodies. To determine the amount on hand, use a figure of 5 pounds of honey per deep frame. Thus, 65 pounds would require thirteen frames. If the requisite amount of honey is not on hand, then feeding is in order. The ratio for fall feeding is two parts sugar to one part water. At this ratio, 10 pounds of sugar should yield about 7 pounds of stores in the hive.

If you determine that feeding is necessary, do it between the end of the nectar flow and the onset of really cold weather. The bees need warm weather to process this food. If they are still flying, it is warm enough for the processing to be done. They do not need to fly to process syrup, but flight is an indicator that the weather is still warm enough. In fall feeding there is no need to simulate a nectar flow as is done for spring feeding. A bulk feeder on top of the hive works well.

Pollen is also a concern in the fall. As a necessary part of the bees' diet, eaten by newly emerged young bees, it must be present in the hive when brood rearing is underway. If no pollen is present, or if none is coming into the hive, brood rearing will not take place. During your fall hive inspections you should look for a minimum of three or four frames of pollen in the hive. The bees will collect and store at least this much for the winter ahead. If you do not see this amount of pollen in the hive as the season is ending, you can do little about it then, but you must be prepared to feed pollen or pollen substitute in the early spring. The bees do not readily accept pollen from the beekeeper in the fall, but they accept it eagerly from any source in early spring.

Hive Protection

A few beekeepers relocate their hives into protected spots for the winter, even moving them inside. Generally, this is not a good idea. I am not talking about indoor wintering in prepared facilities, which is a different concept. I am referring to casual moves into sheds and outbuildings. These are best avoided. Studies have shown that moving hives in the fall may cause bees to winter poorly or consume more food. Your initial selection of a site should take into account any adverse winter conditions. We assume that your hive is in a permanent year-round location.

Windbreaks. Although you chose your site carefully, you still should think about winter winds. If you have any reason to believe

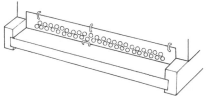

A mouse guard should be placed across the hive entrance in late summer or early fall to prevent mice from establishing winter nests in the hive.

that wind may be excessive and no permanent windbreak exists (a fence or small hedge, for instance), create one. A piece of securely anchored plywood will do, or a couple of bales of hay to break any excessive wind, which otherwise saps energy from the hive.

Mouse guards. Mice find a beehive a wonderful place to spend the winter. They chew out sections of comb and build a nest 4 to 5 inches in diameter, or larger, make a mess, and can cause a colony to abscond. Don't let mice in. Metal is necessary. Mice will chew through a conventional wooden entrance reducer.

It is best to get the mouse guard in place before cold weather arrives. Mice start preparations for winter early, and although the bees usually can keep them out in warm weather, sometimes the mice sneak in on cold nights while the bees are clustered. If you are late in putting on a mouse guard, check first to be sure no mouse has already moved in and will be trapped.

Ventilation and insulation. As has been discussed, during winter, life in the hive continues. Food and oxygen are consumed, and moisture and carbon dioxide are given off. Both of these products of metabolism must be removed from the hive, so ventilation is important. Excess moisture in the hive makes it difficult for the bees to keep warm. Carbon dioxide displaces necessary oxygen. An upper entrance or ventilation port will take care of these problems. A simple solution is to use an insulation board incorporating an upper entrance, above the inner cover. The upper entrance is especially important if snow builds up around the hive and blocks the lower entrance. In

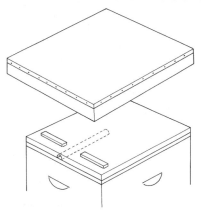

An insulation board, placed over a reversed inner cover (as illustrated), provides an upper entrance and helps keep the hive warm and dry for the winter.

addition to providing insulation, this arrangement will still allow the bees to get out for cleansing trips on warmer winter days.

The illustration shows a system for addressing all of the problems described. At the top, the inner cover is reversed so that the flat side is up. Then an insulating layer goes in place. I use ½-inch thick material called variously insulation board, building board, or wallboard. It is similar to the fibrous material found in ceiling tiles and is available from most building supply sources. The insulating board has a channel on the underside that allows passage for the bees through the hole in the inner cover and out under the edge of the outer cover — an upper entrance. It also allows a flow of moisture and CO_2-laden air out of the hive.

The insulating board serves two purposes. First, there is no air space between it and the inner cover which has been turned upside down. The insulating layer allows the underside of the inner cover to remain warmer than if there were cold air directly above it, as would happen if the inner cover was turned with its rim up. It is then less likely that moisture will condense on the underside of the inner cover. Instead, the moisture, as vapor, will travel with the airflow through the hole in the inner cover and in part will move out of the hive through the ventilating port/upper entrance, and in part will be absorbed by the insulating board, which serves as a wick.

The outer cover is set in place conventionally, but is slightly propped up in front, allowing an airflow across the top of the insulating board. This serves to evaporate out the moisture taken up by the board from below.

Other materials may be used in place of the insulating board. Some beekeepers use a super, without frames, but filled with fiber glass-type insulation. Hay, straw, and leaves have also been used, but these tend to absorb and hold large amounts of moisture, which can defeat the purpose of the insulation. If fibrous material is used, it is best to place a piece of screening so the bees cannot get into it. They often will try to remove it from the hive if they can get at it on warm days.

When the outer cover is slightly propped up as described, it needs a weight on top to keep it securely in place. A few bricks, a rock, or something of similar weight will be fine.

Wrapping hives. Wrapping hives is a different method of hive insulation. It is a practice that was commonly followed by beekeepers in the past, but is not done as often today except perhaps in areas where the winters are extreme. The material most often used has been tar paper, wrapped around the entire hive (or a pair of hives together) and secured with tape or rope. Wrapping a hive is considerably more work than the method described above. Some beekeepers feel that it is worthwhile. This is another instance where you should seek the advice of other beekeepers in your area. Perhaps winters are so severe that wrapping is necessary where you live. A caution, however: be sure to allow an entrance for the bees and upper ventilation.

The Coming Year

In autumn, after you had put your colony to bed for winter, perhaps you gave a small sigh of relief and relaxed a bit, thinking that you had a few months respite. The bees' winter is short, though, and they are active in the depths of winter, on some of the coldest days of the year. Serious brood rearing is underway and the adult population is about to increase dramatically. Large quantities of nectar and pollen will be consumed.

However, if stores are insufficient, the population will not increase and the bees will not live up to their potential. In fact, it is entirely possible for the colony to starve. To preclude this and to ensure that growth takes place, it is necessary to monitor hive activities as the seasons progress into spring. This monitoring may be no more than an occasional external check, or it may require opening the hive for a more detailed inspection. Let's take a closer look at the bees' needs during this period.

Winter

In any part of the country no two winters are precisely the same. On a long-term or seasonal basis, the weather may be warmer or colder than usual. On a short-term basis, there may be greater variations in daily temperatures than is normal — days of exceptionally warm weather or exceptionally cold weather or both. These variations make a difference to the colony. The bees do best in weather that is consistent, and for overall well-being, cooler is better than warmer, since the bees are less active and consume less of their irreplaceable stores when they are relatively inactive.

Excessively cold weather immobilizes the cluster and keeps it small, limiting the activity that can take place within it. Unseasonably warm weather allows the bees to become too active, causing them to use too much of their food reserves. If unseasonably warm weather persists, they may expand the brood nest beyond what can be covered and protected when the colder weather returns.

What this means to the beekeeper is that he or she must pay attention to weather patterns, note the presence or absence of bee activity, and be prepared to help the colony if the situation demands. This help can come in one or two forms. First, routinely check the condition of the hive for snow blocking the entrance or ventilation ports, dead bees blocking entrances, and so on. Then, if all is well in these respects, feeding may be in order if the bees have consumed excess amounts of stores.

Even during the coldest of winters we usually have a warm day or two in February or March or a short spring thaw. On such a warm day check your colony. Are the bees flying? Cleansing flights are important to allow them to void the wastes that have accumulated in their bodies over the past weeks. They are also able to drag out the bodies of bees that have died and fallen to the bottom of the hive.

Check the weight of the hive by lifting the back. If you have done this regularly over the preceding season, and especially after the hive was buttoned up for winter, you should be developing some sense of what it means. Too heavy to move easily? Lots of stores on hand. Lifts easily? Better check inside. Somewhere in between? A judgment call on your part.

If an inside inspection seems necessary, take off the outer cover. Where is the cluster? We know that the cluster establishes itself at the bottom of the hive in the fall, with honey surrounding it on the sides and above. As the winter progresses, the cluster moves slowly upward, consuming the honey to the sides as it works its way to the top. If bees are immediately in view through the hole in the inner cover, then the cluster has worked its way up from below. Stores are getting low. Check now to determine how low.

Remove the inner cover. Use a little smoke. Some novices assume that in the cool winter weather bees cannot or will not fly — they can and they will. Bees from the cluster are warm enough to fly, at least briefly. They can also sting.

With the cover off, check the amount of honey and pollen on hand. You can assume there is no honey below the cluster, but check the frames adjacent to it. Look for a minimum of three to four frames of honey and at least one or two frames of pollen. Actually, you are looking for equivalent amounts. You do not need to find full frames if you can find enough partially filled frames. Proximity of the stores matters, though. A frame of honey that is not near the cluster will not help the bees if the weather is too cold for them to move freely. If you do not find enough honey in the hive, feeding is in order. The form of the feed and method of feeding will depend on the exact time of the season, the weather, and temperature. Feeding sugar syrup in a conventional feeder does not work well in cold weather. When the syrup is cold and the bees are maintaining a cluster, they seldom try to move that syrup from the feeder. If they can extend the cluster to the feeder, or if the day warms up enough, they will use the syrup in daily quantities. When the weather warms up and the bees can move more freely, then they may begin to move the sugar syrup closer.

Feeding syrup. Syrup or honey placed inside the hive works best for cold-weather feeding. A new beekeeper is unlikely to have extra frames of honey on hand, but frames of sugar syrup will work as well. Use a proportion of two to one (sugar to water). Remove a couple of empty frames from the hive, take them to the house, and put the syrup in them. Simply pouring syrup on the frames does not work well; it does not fill the cells properly. Squirt it in under mild pressure. A bottle that once contained liquid dish washing soap will work or pick up an inexpensive spray bottle from a garden supply shop. Use warm (not hot) sugar syrup if possible and put the frames back in the hive while still warm. Place the frames as close to the brood area as possible, without disturbing the integrity of the nest.

Feeding candy. Other methods of feeding sugar do exist. One is to make candy. This is easily done, though it is time consuming. In its simplest form, the candy is made with granulated sugar and water. To make a batch, bring a quart of water to a boil in a medium or large pot. Turn off the heat and add 5 pounds of granulated sugar, stirring constantly. (Five pounds is a lot. You may wish to reduce the amounts of sugar and water proportionately.) When the sugar is completely dissolved, turn on the heat and bring the syrup back to a boil. Keep stirring. Continue boiling until the mixture reaches the hard ball candy stage (260°-270°F), which will take 30 to 40 minutes. At all times throughout this process, be careful not to burn or scorch the sugar. Burned sugar fed to bees can kill them.

Once the mixture has reached hard ball temperature, pour it out on sheets of waxed paper on a flat surface to a depth of ¼ to ⅜ of an inch. As you select your place to pour, keep in mind that this syrup is very hot. It can scorch things. A heavy layer of newspaper under the waxed paper is helpful as insulation.

After the candy has set, it will be hard, somewhat brittle, and light amber in color. Break it into conveniently sized pieces. Although these pieces may be laid on top of the inner cover, it is better to put them under the cover directly on the frames over the cluster.

Feeding dry sugar. Another option for feeding is to use dry granulated sugar, spread on the inner cover. This is strictly an emergency measure to be used when the food reserves are perilously low and syrup or candy are not quickly available. The bees must be able to leave the cluster to get to the sugar and then must have moisture to properly utilize sugar in this form. The bees use it only on a subsistence level and do not attempt to store it. It is often discarded outside the hive when spring arrives and better food becomes available.

Feeding pollen. As a new beekeeper, you are unlikely to have pure bee-collected pollen stored away, but you can obtain pollen substitute. If the bees are short on pollen in winter, use patties made from substitute, or a commercially prepared mix, placed as close to the brood nest as possible. Take a quantity of dry pollen substitute and mix it with enough sugar syrup to make a dough. Form the dough into thin patties, place them between wax paper, and lay them on the frame tops. On top of the brood frames usually works well, but don't place them in such a way as to break up the brood nest.

Pollen patties dry out in the hive. The amount of sugar syrup used in mixing is a balance between making the dough stiff enough to hold shape, but damp enough to remain usable. They can become rock hard. Keep an eye on them and remove and replace them if they do stiffen up. Smaller ones fed more often are helpful in controlling this problem.

Spring

As winter gives way to spring, the bees' needs will increase and your methods to meet those needs can change slightly. During the winter, all may have been fine and you did not need to do anything. As spring progresses though, the rate of brood rearing continues to increase and the rate of food consumption increases accordingly. Keep in mind that not only is population rising, but in warmer weather the bees are more active and are eating more. Food stores may be used up at an alarming rate.

On warm days, even in late winter, you may see signs of foraging — bees actively coming and going, nectar and pollen arriving at the hive. It is encouraging and exciting to see the new year getting underway, but this is really just a preview. Rarely is there enough forage coming in to allow for anything more than stimulation. The bees are not able at this time to put away any reserves. They are probably eating more than they collect, so keep watch and feed as necessary.

When the weather warms up, you have a little more freedom in your methods of feeding both sugar syrup and pollen substitute. A top feeder is probably best for the sugar syrup (a gallon jug on top of the hive). Pollen substitute can be fed in patties as described above, or it may be fed dry outside the hive. If fed dry, place it in a shallow tray in a somewhat obvious place, protected from wind and weather. Offer small amounts and renew it often. The bees will find it quickly and will collect it and take it back to the hive much as they collect natural pollen. However, they often tussle and fight. It is not unusual to see dead bees in the pollen dispenser. Further, you may be feeding bees from other colonies in the neighborhood, not just your own. Therefore, pollen fed as patties inside the hive is usually best.

Getting Ready

When does all of this end? When can you safely conclude that spring is here, feeding can stop, insulation can come off, and you can start thinking about honey surpluses and supers? The answer is tied to your growing season and to the specific bee forage in your area, but it is a simple one. When the first meaningful nectar flow of the season begins, you are off the hook. Obviously, the specific date for this is variable — it is a function of geography and climate. For much of this country, the appearance of dandelions is the signal. If dandelions are not significant in your area, ask other beekeepers what your local signal is. When you get that signal, or when the dandelions begin to bloom, pick a warm pleasant day and proceed with spring chores.

There are four essential spring chores: removal of winter insulation, cleaning inside the hive down to the bottom board, reversing hive bodies, and checking the condition of the bees themselves.

Removing the insulation and cleaning the hive are straightforward operations. If you fed your bees this spring, you may have found it necessary to remove the insulation earlier. If not, remove it now, along with the mouse guard. Then, disassemble the hive, right down to the bottom board. Clean off that bottom board; scrape off any dead bees or other debris. Before replacing the hive bodies, clean them and look the contents over carefully. In their last minute preparations for winter, the bees propolized heavily, sealing up cracks and crevices and filling in empty spaces. They also may have built burr comb in awkward places to increase their honey storage capacity. Scrape that excess propolis and burr comb. Otherwise, you will be coping with it every time you open the hive during the coming season. Between the shoulders of abutting frame end bars is a favorite place for the bees to stuff propolis. You no doubt discovered this during the past season. Be sure to scrape there so the frames will fit more easily. If you don't, it will become progressively more difficult to remove and replace frames. In general, clean up inside the hive so that you will have an easier time working with the bees as the season progresses.

As you are doing this cleanup take the opportunity to observe the contents of the frames. Is brood rearing well underway — six or eight frames of brood as a minimum? Do they show a nice tight brood pattern? If not, perhaps a new queen is in order. If the brood pattern seems good, but the brood nest is small, perhaps the bees got off to a slow start because of a late spring or because of low food reserves. Watch carefully as time passes to see that the population does build. Otherwise, requeening may be in order or you might add a frame or two of brood from another stronger colony if that is available.

Reversing hive bodies. In the fall the bees organized the hive so that the cluster was at the bottom and honey stores were above. The cluster slowly moved up as the winter progressed, eating honey as it went and leaving emptiness below. Now, in the spring, the bees natural inclination is still to work up, but there is no place to go above them. As you replace the hive bodies, if the bottom one is, in fact, completely empty, reverse them and place the empty one on top. Now they have a place above to expand into.

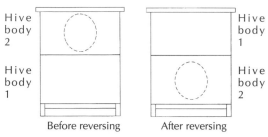

Hive body 2

Hive body 1

Before reversing

Hive body 1

Hive body 2

After reversing

Reversing the hive bodies of this colony in the spring is desirable. No harm will be done to the brood nest and congestion above the nest will be relieved.

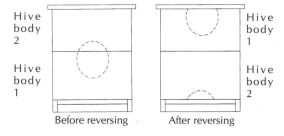

Hive body 2

Hive body 1

Before reversing

Hive body 1

Hive body 2

After reversing

Reversing the hive bodies of this colony is undesirable, since it would split the brood nest. In cool weather, the bees would have difficulty covering the larger surface area of the split nest and brood might die.

If there is brood in both boxes, be careful about splitting the brood nest. There are still cool nights ahead and the colony may not be able to cover a split nest and keep it warm. As an alternative, delay reversing the hive bodies until warmer weather, or move only a few frames, perhaps placing some of the mature capped brood in the bottom box. In this instance, you would place them in the center and consolidate the rest of the brood nest above. Empty frames from below would then be used to fill out at the sides of the top box. There is still only limited space above them, but more than there was, and now there is additional room to the sides. Close up the hive and you and it both should be ready to face the new season.

The New Season

During its first year your colony had certain protections — against disease, against mites, against swarming, even against you. This protection came in part from new equipment, new comb, a well-bred young queen, a certain amount of isolation in a new location, and to a degree, it was protected by your inexperience. Now, during this new season you must come to grips with several new or emphasized aspects of beekeeping — the use of queen excluders, for instance, the techniques of swarm control and prevention, preventing and dealing with diseases and mites, and more. It becomes ever more important to read, to talk with other beekeepers, and to make yourself a part of the larger world of beekeeping. Don't try to go it alone.

CHAPTER 5

REMOVING THE CROP

T wo of the minor mysteries of beekeeping, frequently asked about by nonbeekeepers, concern how to remove the honey from the hive without taking the bees, and later, how to get the honey from the comb. Beekeepers quickly discover that neither operation is difficult, given the necessary knowledge and some equipment. They are two separate operations, of course, with one taking place at the hive and the other indoors. First, we will look at removing the honey from the hive.

Methods for Clearing Bees and Removing Supers

Several choices exist, which variously use brush, bee escape, escape board, fume board, or blower. We will discuss them here in order of simplicity. No matter which method is used, it is helpful to have other equipment handy: first, a spare empty (no frames) hive body or super; second, a piece of heavy cloth, an old towel for instance, large enough to cover a super; third, a means of carrying the supers full of honey. They get heavy. A garden cart is good, a wheelbarrow is okay, and a little red wagon will do.

Bee Brush

The bee brush is a basic piece of beekeeping equipment, used for brushing live bees from frames. Although its usefulness is not restricted to harvest time, that's when it gets the most use. To use it for harvest, first smoke the super gently to get some of the bees moving down and out of the super. After a minute or two, remove the super and set it away from the hive, on the inverted outer cover. Remove the first frame from the super, hold it over the hive (not over the super), and give it one good firm shake. Many of the bees will fall off. Carefully brush the remaining bees into the hive. When the frame is clean of bees, place it in your spare empty super and quickly cover it with the cloth. Use a damp cloth on a breezy day to help keep it in place. The spare super should be in your cart, preferably sitting on a flat surface so bees cannot sneak in under the bottom. The cloth keeps them out at the top.

Do each frame in turn, working as quickly as possible. In the normal course of events, the bees do not get upset with this brushing treatment.

If you do not have a spare super, use any random container that will hold the frames temporarily, such as a cardboard carton. Once the last frame is done, you can brush the active super clean and put the frames back in. In the absence of a box, you can just stack the bee-free frames in your cart, under the cloth, until the last one is done.

Brushing is a simple method. If you have many supers to do, though, it can be tedious, and it can be messy if the brush picks up honey from uncapped cells. You must work carefully so as not to injure bees.

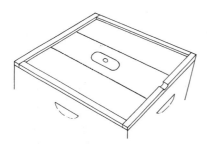

A Porter bee escape, set into the inner cover and under the honey supers, prevents bees from re-entering the supers.

Bee Escape

The Porter bee escape is a one-way exit that is used in conjunction with an inner cover. When placed under the super to be emptied, it allows the bees to pass down through and into the hive body below, but not back into the super again. A minimum of 24 hours is necessary to get all or most of the bees out of the super.

To use the escape, put it into the hole in the inner cover and place the cover between the super and the hive body directly under it. If more than one super is in place, put the escape under the bottom-most super. Close the hive. Allow no bees to enter the super while the escape is in place. Over the 24 hours, the bees inside the supers will move down through the escape to join the rest of the colony below.

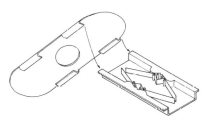

An objection to the Porter bee escape is that often the bees do not use it; they remain in the supers and are waiting there for the beekeeper the next day. There are at least two reasons why this sometimes happens. Either the springs are not properly adjusted

Before using the Porter bee escape, check to see that the springs are properly adjusted and free of debris.

and the bees cannot pass through the escape freely, or there is brood in the supers that the nurse bees will not abandon.

Before using a bee escape, even a brand-new one, check the springs inside. Be sure there is no debris in the springs and that the opening between the tips of the springs at each end is between ⅛ and ³⁄₁₆ of an inch.

Escape Board

The escape board works on the same principle as the Porter bee escape, allowing bees a one-way passage out of the super. In fact, an inner cover with a bee escape in place is an escape board. However, single-purpose boards like the one pictured are available. Other styles are also available. All are used in exactly the same manner as the bee escape. An advan-

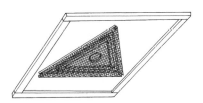

The triangular escape board (underside shown) is another way of emptying bees from the super: they pass through it easily, but are unable to find their way back to the round hole in the center.

tage to the escape board shown is that it is a simple maze, with no springs or other moving parts to become distorted or stuck. The bees pass through it easily but are unable to find their way back.

Fume Board

The fume board uses a foul-smelling liquid to drive bees quickly from the super. It is most effective on a single shallow super in warm weather.

The fume board, sometimes known as the acid board, is used in conjunction with a foul-smelling liquid, the fumes of which drive the bees out of the super. In earlier years, carbolic acid was used commonly (hence the alternate name), but today commercial preparations are available from bee equipment suppliers that are less likely than acid to taint the honey.

To use the fume board, sprinkle a small amount of the liquid on the board, which is then inverted over the super to be emptied. In warm weather, the bees usually vacate the super in 5 to 10 minutes.

Although the fume board works well and quickly under proper conditions, it has some disadvantages. It only works well on warm days. In cooler weather, the fumes are less effective. Furthermore, it does not work as well on deep supers or on more than one shallow super at a time.

Be careful in using and storing the fume board and the bottle of liquid. Do not store them in the house. The odor is foul, strong, and persistent. If you spill any on your clothing, you will not be welcome in polite company.

Blower

A household blower commonly used for clearing fallen leaves is also effective at removing bees from the super.

A blower is a surprisingly effective way to remove bees from supers. It is an expense that may not be justifiable for one or two hives, though. But many households already have a suitable blower — the type used for moving leaves in a yard. Even some of the better shop vacuums may work when operated in reverse, but they require electricity, which is usually not found in the beeyard.

To use a blower, first remove the super and stand it on one end next to the hive. A stand or platform is helpful; an empty hive body works. Position the super with the bottom toward you so that the bees will be blown toward the front of the hive. Direct the air flow up and down and back and forth between the frames until all or at least most of the bees are out.

A super with lots of burr and bridge comb is difficult to clear. The bees hide behind the obstruc-

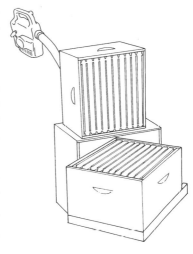

When blowing bees from the super, place the super on its end (with the bottom toward you) and blow the bees toward the front of the hive. Direct the air flow between the frames.

tions. Sometimes turning the super and blowing from the top does the job.

None of the methods described above is foolproof. For any of the last four, you may find that you must resort to a brush to get the last few bees out. Be prepared for this and always have a brush with you.

What to Do Next

You *must* protect the honey once you have the supers off the hive — no matter what method you use for clearing the bees. The degree and nature of the protection depends on how long it will be before you actually extract the honey from the supers. Your first concern is to put the supers under cover so the bees cannot recover their wealth. The process of removing honey from the hives has filled the air with its odor and has no doubt agitated the bees. Some of them will follow the supers and will attempt to steal it back. If you give them any kind of opportunity, they will exploit it and you may lose the crop. It is amazing how much of that honey they are able to squeeze back into the hive, even with no supers in place, once robbing gets underway. And of course, once thievery starts, you can expect other colonies in the neighborhood to join in.

Your first concern, then, is to put the honey inside, preferably in a location where the bees cannot even smell it. If they know it is there — especially if the nectar flow is over — they will hang around continually, trying to get in.

Your second concern is temperature and humidity. Honey flows best when it is warm; cold honey is more difficult to extract. If the honey is being taken from the hive toward the end of the season, before really cold weather sets in, you should be able to extract with little difficulty. To make your job easier, keep the supers at room temperature or higher until you actually do extract.

Honey is hygroscopic: it absorbs and retains moisture. The natural environment inside a healthy hive is such that there is no moisture problem. Once off the hive though, the honey is no longer protected by the bees and may take up atmospheric moisture. High-moisture honey can ferment. Do not be overly concerned about this, however. A day or two's delay is not going to affect the moisture content appreciably. If you must store it for a longer time, though, a dehumidifier in the same room helps maintain or even lower the moisture content.

Be aware that the supers will drip. The process of removing them from the hive has unavoidably broken some of the burr and bridge comb, so that honey oozes and runs out onto the floor. You will need something to catch it. A spare outer cover, inverted, makes a suitable tray.

Now that the honey is off the hive and under cover, we can get on with the extraction process.

Extracting and Handling Honey

Extracting — removing the honey from the supers — is your last big chore of the season. It is a relatively straightforward task, but it has its nuances. The best way to remove honey from the supers, and just about the only way, is to use an extractor, a piece of equipment that whirls the frames and removes the honey by centrifugal force.

One alternative to an extractor does exist, although it is not a good one: removing the honey-laden comb from the frames and crushing it, either in a mesh bag or in a large strainer. You can make a mesh bag from cheesecloth, nylon, or any other material that will hold back the broken up comb while allowing the honey to strain through. It is a messy process and takes time — 24 hours or longer per load — unless you use a press of some sort to squeeze out the honey. It helps to massage the bag and its contents periodically as it drains to encourage the process.

Instead of a mesh bag, you could make or acquire a strainer, such as an oversized colander. It might be necessary to use the mesh material in conjunction with this if the holes in the strainer are large. Otherwise, the honey will be full of particles of wax and other debris.

A big disadvantage of the crushing method, aside from the messiness of the process, is that the comb is destroyed. The bees must start off with new foundation in the following season. Furthermore, it is not a practical method when any significant amount of honey is involved. Whichever method you use, a pot or a pail to catch the drippings is necessary.

Most beekeepers do manage to find an extractor. They are expensive, though — prices start at over $200 for a small one. You could share the cost, of course, and buy one cooperatively with one or more other beekeepers. Another possibility is to build one. This can be done without much difficulty for someone so inclined; plans for building a small extractor inexpensively do exist. Another possibility is to rent one. Many beekeeping equipment suppliers keep a rental unit on hand, so ask around.

Assuming that you have located an extractor, what else do you need? Several things. First, a place to work. Extracting is messy. Even processing a small amount of honey can spread stickiness over a large area. In the early days I extracted in the kitchen. This is no problem, but be prepared for lots of cleanup. Maybe it won't be that bad for you — maybe you are a neat worker.

Besides a place, you should allow plenty of time. If you decide to do it in the kitchen, don't try to sneak it in between 3 o'clock and dinner. You won't be finished in time, especially during your first attempt when you are not really sure of what you are doing. Wait until evening.

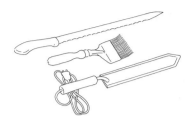

Tools useful for uncapping frames include (top to bottom): a kitchen carving knife, a capping scratcher, and an electric knife.

Additional Equipment

Some additional equipment is necessary — uncapping knife, capping scratcher, cappings container, strainer, pails, honey gates, bottling tank. Let's discuss each one.

Uncapping knife. Some kind of blade is necessary to remove the wax cappings from the comb before you can extract. The beekeeping catalogs show several uncapping knives and a plane. They are relatively expensive, even for the nonelectric versions. Consider using one of your kitchen knives the first time you extract. For one or two supers, it will be adequate. A serrated bread knife or carving knife works well, and if the blade is warmed in a pan of water periodically, it will work better.

If you are ready for something more sophisticated, your choice is between an electric knife and an electric plane. My preference is for the knife. I find the plane too heavy and awkward, but many beekeepers prefer it.

Capping scratcher. This handy tool gets the low spots. A frame of honey, especially one newly drawn and filled, may not be drawn to a uniform depth. The uncapping knife, as it is moves across the frame, may miss some of the cappings. The scratcher can be dragged across these missed cappings to break them up.

Cappings container. As you uncap, you will need a container, a tub or pail, to catch the cappings. They tend to flop as they fall free from the comb, so a wide container is best. A large amount of excellent honey adheres to these cappings. The container should ideally have a strainer near the bottom for the cappings to rest on and a gate or valve at the low point for removing that honey after it drains through.

Strainer. As the honey comes from the extractor, it contains particles of wax, bits of propolis, occasional wings or legs, and other hive debris. You have a choice. You may strain it through a coarse or medium mesh of cloth or metal as it comes from the extractor. Alternatively, you can run it directly into a pail or other holding container and leave it for 24 hours or more in a warm room. At the end of that time the impurities will have risen to the top and can be skimmed off. After either coarse straining or skimming, you may wish to strain the honey through a filter cloth (70-mesh nylon, for instance) to remove any very fine particles remaining.

Plastic pails. A beekeeper never seems to have enough pails. The 5-gallon size is most commonly used, with the 4-gallon running second. Larger pails are available, but honey is heavy and the larger ones can be awkward to handle. Pails can be purchased from some bee supply dealers and from other sources, but used ones are often available free or at small

Following extraction, the honey may be strained to remove coarse particles of wax, propolis, and other hive debris.

cost from bakeries and restaurants. Use *only* food-quality containers — those that were manufactured to contain food and have been used exclusively for food. Food-quality pails are usually white, but being white does not guarantee their quality. Be sure to inquire about their past use.

Honey gates. In beekeeper parlance, a gate is another name for the valve or faucet used in honey containers. Honey is a thick and viscous material. It needs a valve that shuts off the flow quickly and cleanly with a minimum of dripping. Honey gates do this. Gates are available in two materials, plastic and brass. Brass is normally used for the larger heavy-duty applications. For most small operations, plastic serves well.

Bottling tank. Once you have your honey out of the extractor, strained, and settled, it is ready for bottling. You can move it directly from a pail into a jar, but only with considerable difficulty and probably with lots

You can make your own honey bottling tank with a 5- or 6-gallon plastic pail and a threaded, plastic honey gate.

of spillage. Because of its viscosity, honey does not pour easily. You need a means to cut off the flow sharply — a bottling tank with a honey gate should, therefore, be considered essential. You can buy a small tank, already made, or you can make one yourself. A 5- or 6-gallon plastic pail with a 1¼-inch, threaded, plastic honey gate are all that are needed. Cut a hole low in the side of the pail, insert the gate, screw the nut on the back, and you're done. (When you cut the hole, be sure to leave room on the inside for the nut to turn.)

The Extracting Process

Extractors come in various sizes, measured by the number of frames they will accommodate at one time. Beginners most often use an extractor of two-, three-, or four-frame capacity. Most extractors have a reservoir below the basket where honey can accumulate as you proceed, with a honey gate to drain the reservoir. Depending on the design of the particular extractor, you may wish to keep the gate open so that honey drains through as you work. This way, it will not build up around the bottom of the basket and impede the rotation. However, this means you must have the extractor up on a platform so a pail will fit under the gate.

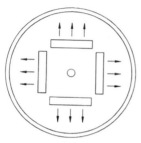

The tangential style is most common for small extractors. Frames are positioned as illustrated, with honey spun from just one side of each frame at a time.

Perhaps your extractor will have an ample reservoir at the bottom and you can allow it to sit on the floor with the gate closed, raising and draining it periodically as you proceed.

Assume that the extractor will vibrate a bit and will try to walk about the room as you work, especially if the load is not balanced. Fasten it down somehow before you start or have a helper hold the extractor in place as you spin. Balance the load as best you can by putting frames of like weight in the same load. If you come to the end of the job without a full load of frames, use previously extracted empty frames to partially balance things.

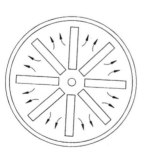

In a radial extractor, honey is spun from both sides of the frames at once.

Extractors come in two styles —

USING A TANGENTIAL EXTRACTOR

To use a tangential extractor, load up and spin out an estimated one-half of the honey from the first side. Then, remove, reverse, and replace those frames. Spin out all the honey from the second side. Reverse and replace again to finish on the first side. This is a bit of extra handling, but it helps to ensure that the weight of honey on the second side does not push through and break the somewhat tender comb of a new frame. The bees strengthen the comb a little more each year so in subsequent years the comb is not quite so fragile, but a little care in this respect will always be in order.

tangential and radial. Most of the small extractors are tangential, which means the frames are positioned at an angle to a radius of the extractor tank. This is the style you are most likely to use first. Later, if you expand your operation, you may acquire a radial extractor. In a radial extractor, the frames are positioned along a radius. The difference is that in a tangential extractor, honey is spun from only one side of each frame. In the radial type, honey is spun from both sides at once.

You have no doubt heard the expression "slower than cold molasses." Honey is the same — perhaps even slower. If you are able to store the supers in a warm room for a day or so before you extract, the honey comes from the frames much more easily and passes through your strainers more quickly.

As you process the honey, moving it from one container to another, try to minimize splashing to reduce the possibility of bubbles and foaming. Each step of the extracting, straining, and bottling process has the potential to add air to the honey. Air, of course, is not harmful, but it can be unsightly once the honey is in the jar. Also, allow the honey to sit in the bottling tank at room temperature for at least a few hours, preferably for a day or two. The bubbles rise and can be skimmed. Then bottling through a gate at the bottom should give nice clear honey in the jar.

If you have only a small quantity of honey from your extracting process, you probably will bottle it as a part of that process. However, if you have had a banner year, perhaps you will be storing some away. In that case, maybe you will choose not to bottle it now.

Storing Your Honey

Honey granulates. This is a fact of beekeeping life. Granulation (or crystallization) is a natural state of most honeys. It happens as a function of time and of the particular combination of sugars in the honey. Assume that any or all of your honey may granulate at some time or another. The honey is in no way damaged by this natural process and it can be returned to its liquid state with relative ease. A defense against granulation is heat. Judiciously applied, heat can stop or delay the process by dissolving and destroying the microscopic seed crystals that are naturally found in honey. However, heat can damage honey.

Large commercial honey producers and packers apply heat during the extracting/packing process and their honey is protected against granulation almost indefinitely. This heating is done under controlled conditions. The honey is heated and cooled rapidly with temperatures reaching 160°F (71°C) in the process. Most small operators do not have the capability to heat and then cool bulk honey rapidly, and honey can be damaged if left at 160°F (71°C) until it cools naturally. Honey can be heated to lesser temperatures that will protect it for shorter periods of time.

Actually, any amount of heat causes some deterioration in the quality of honey, although to most of us it is not noticeable. Beekeepers accept this deterioration and work to minimize it. Toward this end, as a small operator, never heat honey above 120°F (49°C) and do not heat it any more often than absolutely necessary.

Honey in bulk is slower to granulate than honey in small jars. Honey in a 5-gallon pail may stay liquid for months while a 1-pound jar filled from that same pail may granulate in weeks. Because of this, you may wish to store honey in bulk if you have no immediate need for it. Consider using containers of at least 1-gallon capacity as your storage media rather than smaller jars. If the honey does granulate in storage, 1-gallon containers can be handled with relative ease when reliquifying the honey.

To reliquify honey, no matter what the container size, always use a double boiler arrangement. Place the container of honey in a pot of water that will allow the water to rise high around the honey. Bring the water temperature to about 120°F (49°C). Maintain that temperature by reheating periodically until the honey returns to its liquid state.

The optimum temperature for honey granulation is about 57°F (14°C). To avoid or delay granulation when storing honey, room temperature is best. Refrigeration will hasten the process. Honey may be frozen with no ill effects and frozen honey will not granulate — a few hours at room temperature and it is ready for use.

Your one remaining task now is to find a use for all of that honey. I'm confident that you will.

Comb Honey

Because this book is primarily for beginners — new beekeepers with new colonies — an implicit assumption has been made that the surplus honey crop in the first year will be small, perhaps even nonexistent. Earlier, we considered optimistically what to do with a honey surplus should it exist.

Some experienced beekeepers recommend to beginners that in order to avoid the hassle of extracting, they not even attempt to make a crop of extracted honey in their first year, but instead concentrate on comb honey. I think this is folly. I strongly recommend that as a beginner you not be carried away by the expense, the work, and the disappointment that is almost certain to result from attempting to make comb honey with a new colony.

Making comb honey is a specialized operation that requires knowledge and understanding of the bees beyond that of most first-year beekeepers. It also requires a good population of bees in the early season, which is not a normal attribute of a new colony. Supers equipped for comb honey production can be placed on a colony in midsummer and later, but the results are almost always disappointing. Sections may not be completely filled or properly capped, the wax may be too tough for enjoyable eating, and in the late season, excess propolis on or in the wax may be a problem. In spite of this, let me now tell you how you may be able to enjoy a little (even a lot of) comb honey in your first season, not as an alternative to extracted honey, but in conjunction with it.

This assumes, of course, that your colony has built up to the point where you are ready to put a super on for surplus honey. Do set up your super for extracted honey, but with only seven or eight frames of wired foundation. The remaining two or three frames can then be set up with comb foundation, pinned in place. Comb foundation is manufactured specifically for the purpose. It is thinner, whiter, and has no embedded wire. Ultimately, it will be eaten, so handle it accordingly. Place these latter frames in the middle of the super. This is where the bees are most likely to work first.

If all goes well, the colony will draw and fill the comb and you will have some acceptable sections. If you were overly optimistic and they weren't truly ready, it is no big loss. You still have those prepared frames. Perhaps the bees did not touch them. Put them away for another year. Perhaps they are partially drawn and even have some ripe honey in them, though not acceptable as comb honey. In that case, extract it. Be careful with the comb in doing so; it is not as strong as wired extracting comb. Next year, give these frames back as extracting comb. The bees will strengthen it with wax and propolis and in 2 or 3 years you may have difficulty distinguishing it from regular wired comb.

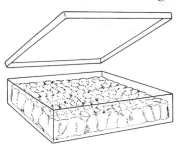

Cut comb honey can be conveniently packed in square plastic boxes, available from bee equipment suppliers.

If all went well and you do have some acceptable comb honey, cut it carefully from the frames with a warm knife. Some cells will be cut open in the process and, as a result, honey will run. Allow the sections to drain on a wire rack over a shallow pan for a few hours in a warm room, then pack it in boxes. Draining is not absolutely required; it is primarily for appearance. Otherwise the comb section would sit in the box in an unsightly puddle.

Use square plastic boxes, available from bee equipment suppliers, to pack comb honey. A shallow frame, completely filled, will yield four equal sections of comb that fit these boxes perfectly. If the frames are not completely filled, cut out the best parts and put the scraps in a bowl. Leave it out to snack on — it's delicious.

The question often arises about how to use comb honey. It is often eaten directly, by the spoonful, as a substitute for candy. The wax is edible. It can be chewed up and swallowed or treated like gum — once the goodness is chewed out, dispose of it. Comb honey can also be spread on toast or muffins, wax and all.

If you have only a modest number of sections, your family will use them quickly. Otherwise, they make wonderful gifts. And a gift of comb honey is a good way to introduce people to one of the pleasures of beekeeping, especially those neighbors who may be a little apprehensive about your bees.

For long-term storage, place the sections of comb in a plastic bag in the freezer so they do not granulate. However, even granulated, it is good to eat.

Other Hive Products

During your first year or two as a beekeeper, you should not be overly concerned with other hive products, but be aware of them. They are of interest to many people and you will probably be questioned. The five other products of the hive that are frequently discussed (and sometimes collected) are beeswax, pollen, propolis, royal jelly, and venom. Two of these have further uses by beekeepers, but other uses exist for all five, primarily as natural food or health supplements. The efficacy of some of these health uses is not conclusive.

Beeswax

Beeswax has a variety of uses. A very large quantity is recycled to become the beeswax foundation we use in our hives. It is commonly used in candles, either pure or mixed with paraffin; it is often an ingredient in certain types of cosmetics, creams, and salves; and

APITHERAPY

Apitherapy, sometimes written apiotherapy, is the collective term used to cover any aspect of health treatment involving products of the hive. As such, it covers all the products discussed here, including honey. This is a growing field of interest and study, and though much of the medical community does not agree, it is a field that holds promise. At least one organization exists in the United States to further this work — the American Apitherapy Society, P.O. Box 74, North Hartland, VT 05052.

it is sometimes a component in wood and metal polishes.

Over time, a beekeeper tends to accumulate significant quantities of wax, which come from three basic sources: cappings from the honey extracting process, scrapings of burr and bridge comb from working the hive, and melted down older comb. As a new beekeeper, you are not likely to have much, if any, of the latter. Presumably, your comb is new and will remain in good condition for at least 3 or 4 years. After that you will start to recycle some of the comb as it ages and deteriorates.

Cappings and scrapings, when melted down and cleaned, usually yield high-quality, light-colored wax. This should be kept separate from melted-down brood comb, which is almost certainly dark and contains impurities not easily removed by the beekeeper. The lighter waxes command premium prices when sold and are the only acceptable wax for use in candles and salves, for instance. The darker waxes can be treated in industrial processes to bleach their color, but this is not something a small operator can easily do.

Pollen

Pollen is often collected at the hive by beekeepers and fed back to the bees in times of need, or sold as a health food supplement. Special pollen-trapping apparatus is available from bee supply sources. During its first season, a colony should not be subjected to the stresses of pollen trapping. Later, you may choose to do so, but it is a subject to be researched carefully first.

Pollen is often touted as a perfect food for humans because it is said to contain all of the vitamins and the minerals necessary for human health. Users frequently extol its benefits. No one seriously questions whether pollen does in fact contain these vitamins and minerals. There is a question, however, as to whether the human body is capable of breaking down pollen and using it.

A further concern is that we are often not aware of the source of pollen offered for sale. There is no convenient way to know if the pollen was collected from plants recently treated with pesticides. This can be a concern even with your own bees. You are never sure where they have been foraging.

Propolis

Propolis is a resinous substance collected by the bees from the bark and buds of trees and other plant materials. It is used in the hive to seal up cracks and crevices, making the hive weathertight. It also is used as a component of brood comb to give it added strength.

Propolis is heavily used late in the season as the bees prepare for winter, but it is found in the hive in any season. In warm weather it is sticky and tenacious. In cool weather it becomes hard and often brittle. It is a nuisance to the beekeeper, but important to the bees.

Propolis is often collected, cleaned, and used as a natural health aid, such as an ingredient in toothpaste. It is also considered to be of benefit as a tincture in treating wounds. As with pollen, there is very limited scientific evidence for its value to humans, although anecdotal evidence abounds.

Royal Jelly

Royal jelly, sometimes called bee milk, is a secretion from glands in a worker bee's head. Its specific value to bees is as brood food, fed to every larva during its first three days of life. A developing queen receives royal jelly for her entire larval life. To this diet, plus the size of the cell in which she develops, is attributed everything that makes a queen different from a worker.

Humans have extrapolated from this benefit to the queen that there must be benefits to humans, and royal jelly is collected and offered for sale as a health supplement. Again, there is little, if any, scientific evidence of such value.

Bee Venom

Bee venom has a definite value to humans: it is collected and used medically to desensitize individuals who have serious allergic reactions to bee stings. As such, it is sought after by drug companies. However, the volume of venom they need is very small and the market channels are established. No real financial opportunities exist at this time for an individual beekeeper who might be interested in collecting and selling venom.

Bee venom has also been demonstrated to benefit individuals suffering from certain forms of arthritis and there is some evidence that it may be of benefit in treating other human ailments. Some research is underway, although much of the present evidence is anecdotal. Aside from treating sting allergies, the health benefits of bee venom are not widely accepted by the medical community. Most of the work being done in this area uses live bees to inject venom.

CHAPTER 6

DISEASES, PESTS, AND PREDATORS

H oney bees are subject to disease. Some are minor and no worse for the bees than a common cold to most of us. Others are life threatening, putting the existence of the entire colony at risk. These diseases come in several forms: bacteria, viruses, protozoa, and fungi. Some of them attack the brood, some the adults. No bee disease is known to harm humans in any way and we may eat honey from an infected colony without qualms.

Diseases of Honey Bees

These fall into two groups: those that affect adult bees and those that affect brood. Generally, brood diseases are the most devastating to a colony, though adult disease can have serious effects. All of the diseases can be treated to various degrees with either medication or specific hive management techniques.

One of the advantages of starting a colony from a package is that these bees are much less likely to become diseased during their first season if hived in new equipment. But disease is still a possibility, so a new beekeeper should be aware of it and prepared in the event that his or her colony is afflicted.

Five diseases (four brood diseases and one adult bee disease) are of particular concern. Others do exist, but are rarely seen or recognized by the average beekeeper. Each of the five are described as are the various methods of treatment.

If you encounter a condition in your hive that you are unable to diagnose, contact your state apiary inspector. In the absence of an inspector, find a competent experienced beekeeper to give you an opinion. A persistent, puzzling condition in a colony should not be ignored. It may get worse — much worse.

American Foulbrood

American foulbrood (AFB), the most serious of the brood diseases, is the reason that state apiary inspection programs came into being many years ago. AFB is caused by spore-forming bacteria. It affects brood only, though it is carried by adult bees, primarily through the medium of honey. The spores are also found in the equipment of a diseased colony, particularly in the comb, whether bees are also present or not, and these spores can persist for 30 to 40 years at least.

When AFB is not treated, it is deadly to a colony. The brood dies, population dwindles, and soon the colony dies out. Other colonies in the neighborhood discover this dead colony, investigate, help themselves to any remaining stored honey, take AFB home to infect their own colony, and the cycle is repeated. Foul brood in one colony in a beeyard, unchecked, is almost certainly going to infect all of the colonies in that yard eventually, and probably any other colonies in the neighborhood. It cannot be left undiscovered and untreated. Hence, inspection programs; but we should not be totally dependent on others. The beekeeper's first defense against AFB or any other bee disease is a thorough personal understanding of the disease: its appearance, its effects, and its treatment.

AFB affects brood only. Specifically, the spores are carried in brood food, which infect larvae. The spores germinate and multiply in the larval gut and the larvae die. Usually, at the time of death, the individual larva has already stretched out and spun its cocoon, and the cell has been capped. In the cell the larva or propupa turns brown and begins to putrefy, giving off the typical sour, dirty-sock or glue-pot odor so often associated with this disease. The cappings darken and sink in slightly, taking on a shiny unnatural appearance when viewed as a mass. The workers begin to uncap these cells, first making small holes in the cappings, later removing them entirely.

When the cells are uncapped, the larvae will have begun to "melt down." They typically have a coffee-brown color and a semiliquid consistency. At this point, a toothpick thrust into the gooey mass and withdrawn will bring out with it a ropy thread, usually an inch or more in length. Later, this material dries out and forms a black scale which adheres tightly to the bottom wall of the cell. The scale is almost impossible for either bees or beekeepers to remove.

During the semiliquid stage, the bees attempt to remove the remains. However, as time passes, more brood become infected and die, fewer brood are surviving to emerge as adults, and the hive population dwindles as adult bees die naturally and are not replaced. Soon the cleanup work to be done far outpaces the available bees and the spore-laden dead brood remain to contaminate the comb. Ultimately, the colony dies out.

Prevention and treatment. American foulbrood can be prevented, and if discovered in the early stages of colony infection, it can be treated. However, some states prohibit treatment and require that colonies with AFB be destroyed, no matter what the level of infection.

Prevention and treatment are accomplished by treating each hive with an antibiotic. Presently there is only one antibiotic specifically labeled and allowed for use in a beehive in the United States: Terramycin (oxytetracycline hydrochloride). Terramycin is available in several formulations and some of these are commonly used in treating animals other than bees. Some of these formulations may not be totally effective for bees. To be sure you are obtaining a proper formulation, it is best to buy it from a bee supply house. As with any medication, use only in accordance with the label instructions for honey bees.

Terramycin is effective in preventing AFB if used regularly. The treatment does not actually kill the spores, but it does enable the brood to withstand the infection. With treatment, the spores can be present in the colony and not affect it. This means that treatment, once started, must continue indefinitely. There is no practical and simple way to know if spores have come into a previously uninfected colony that has been undergoing routine preventive treatment with Terramycin. For complete protection, treatment is administered twice per year, once, before honey supers are put on in the spring, and again after they have been removed in the fall. No medication of any kind should be administered while honey supers are in place. This means that if disease is discovered in a colony during the active season, supers must come off before treatment begins. Further, the amount of time between the end of treatment and the placement of supers is specific. Consult the medication label instructions for details.

If AFB is detected in a colony, treatment should begin immediately if allowed in your state. However, at some point, the disease will be too far advanced for treatment to be effective. It may be best to kill the bees immediately and clean up the equipment. My own rule of thumb is that if there are only a few cells showing AFB in only one or two frames, it is treatable. Beyond that, I do at least consider destroying the colony.

Handling diseased equipment. If a colony has to be destroyed because of American foulbrood infection, the method usually prescribed involves burning. Some beekeepers burn the entire hive, others burn only the frames and comb and sterilize the remainder. I favor the latter approach, but either procedure must be done with thought and care. There is no point in destroying the bees and equipment if all the sources of infection are not totally removed.

The bees must first be killed if a hive has to be destroyed. Otherwise, at least some of them may move into another colony and carry infection there. This destruction may be accomplished by first sliding a pan with several ounces of gasoline into the hive entrance, then sealing up all openings to the hive. Alternatively, the gasoline may be poured into the top of the hive, over the frames, after which the hive is closed tightly. The fumes will kill the bees.

Choose a place where you can dig a pit in the ground, one that is unlikely to be disturbed for some time to come. After the bees are dead, burn at least the frames, the comb, and the dead bees in the pit. First build a hot fire, then place the material to be burned on the fire. Be sure everything is thoroughly consumed by flames. Cover the pit carefully with dirt once the coals have cooled.

Hive bodies, supers, bottoms, and covers may be sterilized and reused. First, scrape everything carefully, using your hive tool. Dispose of the scrapings, preferably by burning in the pit. Then use a propane torch to scorch all of the inner surfaces of the hive parts. It is not necessary to actually burn or char the wood, just a gentle scorching. The equipment should then be suitable for reuse. Sterilize your hive tool with the torch, when you are finished.

Other methods for sterilizing equipment do exist. Boiling the equipment in lye water or placing it in a radiation chamber are two. These methods have drawbacks however. Radiation is not always available, and lye is dangerous.

European Foulbrood

European foulbrood (EFB) is generally considered less serious a disease than American foulbrood; it often clears up spontaneously. However, it is of concern, and alone or in combination with another disease can cause a colony to die. EFB has certain similarities to AFB and is also caused by a bacterium, but it is different.

Larvae are infected with EFB through contaminated food and most infected larvae die, usually before cell capping. The dead larvae are slightly displaced in their cells, become discolored, and begin to decompose. As decomposition proceeds, a toothpick thrust into and withdrawn from the remains can sometimes result in a short, ropy thread (perhaps ¼ to ½ inch long). This should not be confused with the longer thread of AFB. There also may be a slightly sour odor, although it is not always detectable and is not as distinctive as the odor associated with AFB. And finally, the larval remains do not adhere to the cell walls as tenaciously as do those of AFB. The bees are able to remove them if the worker population is not too badly reduced.

Treatment. Colonies suffering from EFB often recover spontaneously, but other diseases and stresses can hinder or prevent recovery. A first line of defense is to maintain all colonies in robust good health so they can withstand the disease. A colony that contracts EFB can often be helped by the addition of a frame or two of capped brood from another colony. This gives the diseased colony a quick boost in its young bee population, adding new workers to clean up the hive.

Another possibility is to requeen with a young, vigorous queen. The mechanics of requeening will cause a slight break in brood rearing, allowing the existing work force more opportunity to clean up diseased larvae. Terramycin has also been found to help, although not to the same degree as with AFB. As a last resort, burning may be the only answer for severely infected hives. The same criteria and procedure applies here as with AFB.

Sacbrood

Sacbrood is caused by a virus and most colonies carry the disease organisms. It is generally considered a stress disease, not overly serious in itself, but it can cause losses when the hive is otherwise weakened.

Infected larvae die after the cells have been capped. They fail to spin a cocoon and the larval skin toughens to form a sac (giving the disease its name) inside of which watery secretions accumulate. The workers uncap the diseased cells to reveal this sac, which usually is yellowish, changing later to dark brown. If the larva is not removed, it dries down to a flattened dark scale. Before it dries out, the sac can easily be removed from the cell by reaching in carefully with a toothpick, pulling it out intact.

Treatment. No known chemical treatment exists for virus infections of bees. Terramycin is not effective. As with other stress diseases, the first and best treatment is to keep the colony strong so that the bees can keep ahead of the disease by cleaning out diseased larvae as soon as they are detected. Colonies showing a heavy infection of sacbrood can be treated as for EFB, by requeening or by adding frames of capped brood from a healthy colony.

Chalkbrood

Chalkbrood is a fungal disease that infects larvae through spores ingested with their food. The larvae die soon after the cells are capped. After the cells are uncapped, the larvae appear as hard, whitish (or sometimes grayish), chalky lumps. These remains are loose in the cells and can be shaken out. They are often seen on the bottom board or at the entrance where the bees have dropped them as they attempt to remove them from the hive.

Chalkbrood is relatively new in North America but has become widespread. Although it is not generally considered serious, it can be detrimental to colonies, with large areas of infected brood sometimes showing up on one or several frames.

Treatment. A mild infection of a few hive cells is not uncommon and usually can be handled by the bees. The beekeeper can help by gently shaking out the chalky lumps, away from the hive. However, it is best to remove and burn frames showing large amounts of chalkbrood.

No chemical treatments are available for chalkbrood. As with EFB and sacbrood, requeening or an infusion of young bees from a healthy colony may help.

Nosema Disease

Nosema disease is caused by a protozoa. It is an infection of the gut of adult bees — workers, drones, and the queen. The disease is widespread and serious. Its effects on workers are a less productive life and premature death. Its effects on the queen are first, reduced egg laying, and ultimately perhaps, her death. The effect on the colony, of course, is a weakened and dwindling population.

Nosema disease has its most devastating effects on a colony in the late winter, a period when colonies are otherwise already stressed and population is normally at its lowest point for the year.

There are no positive outward symptoms of nosema. Infected adult bees show no signs of the disease and a symptom sometimes attributed to nosema — spotting of the hive with feces — can have other causes. The only positive diagnosis comes from dissection and microscopic examination of adult bees or of their fecal matter. Because nosema is widespread in North America and is serious, it is probably best to assume that all colonies have it and treat them accordingly.

Treatment. Nosema disease may be treated with fumagillin, which is available under the trade name Fumidil B. This medication is administered twice a year, spring and fall, in sugar syrup. Fumagillin does not eliminate the infection, but brings it under control so that the colony can thrive. The twice yearly treatment should be routine.

Nosema and mites. Routine treatment for nosema disease has become even more important in recent years with the arrival in North America of tracheal mites, which are also widespread. Their debilitating effects on an infected colony are similar to the effects of nosema. Taken together, the results are devastating, so treat your colony routinely to protect against both.

The foregoing has been only a brief introduction to some of the more common and more serious diseases of honey bees. A great deal more information is available. A beekeeper's methods and techniques for hive management can have a great affect on the impact of disease on his or her colony. It will benefit you and your bees for you to read more on these and other diseases.

Parasitic Mites

During the 1980s, two different parasitic mites of honey bees came into North America, causing untold problems for bees and for beekeepers alike. One of these, the honey bee tracheal mite, *Acarapis woodii*, is an internal parasite of the honey bee. It lives and breeds in the bees' tracheal (breathing) tubes and feeds on hemolymph (blood). The Varroa mite, *Varroa jacobsonii,* is an external parasite that also feeds on hemolymph of both adult bees and brood. The entire life cycle of these mites occurs with the bees.

These mites have become a very serious problem in the United States, causing the loss of thousands of colonies of honey bees. Although no means has yet been found to completely eliminate them, a measure of control is possible with proper treatment.

The Honey Bee Tracheal Mite

Tracheal mites were first identified in England in 1921. Some authorities believe they evolved there from another mite, emerging as a new species in the early part of the century. Those that came into this country were apparently from sources on two continents, Mexico and Europe. There are differences in the two mites, suggesting two separate races.

Since its arrival in the United States the tracheal mite has spread throughout the country by two primary means — by migratory beekeepers and through distribution of package bees and queens. Although tracheal mites are now thought to be found everywhere, not every colony or yard is necessarily infested.

Some of the earlier reports on tracheal mites suggested that the problem in this country would not be severe. This supposition was based on experience in other parts of the world, especially in England. Experience here has shown the problem to be very severe, with a net loss nationally since the late 1980s of at least several hundred thousand colonies. This mite has now been accepted as a fact of beekeeping life, not controllable on a national or regional level, so less attention is being given to it by regulatory agencies. It has been left to individual beekeepers to recognize and control tracheal mites in their own holdings.

Life cycle of the tracheal mite. The female mite enters the tracheal tube of an adult honey bee through the first thoracic spiracle and takes up residence. She lays eggs that hatch, develop, and become adult mites within the trachea. These new adults are both male and female. They mate, after which the new females leave the host bee to find another host to begin laying eggs. The time from egg to egg, the complete life cycle, is 14 days.

Transfer to the new host takes place after the mite exits the tracheal tube through the spiracle. After exiting, she goes to the end of one of the surrounding hairs and waits for a passing bee as bees circulate in the hive. Not every adult bee is acceptable as a new host: the mite looks for a young bee.

Within the bee's tracheae, mites in all stages of development are found: eggs, larvae, and adult. The tracheal tubes can become very cluttered and eventually blocked. However, it is not clear that this blockage is what kills the bees; the flow of oxygen does not appear to be inhibited. Infested bees live and work normally, although their lives are shortened. The precise cause of death is uncertain.

The mite population is cyclical. There is a fall buildup, a winter peak, and a summer crash. This is opposite to the normal honey bee population cycle in which there is a spring buildup, a summer peak, and a winter decline. A good queen bee will outproduce the mites in spring and summer, getting ahead of them and easing the mite problem for the colony. Spring requeening helps boost the bee population further, and menthol treatment in the late season helps to control the mite's winter buildup. The greatest losses of both individual bees and of colonies are in the late winter and early spring. This is not surprising considering the mite's winter buildup and the bees' normal difficulties in trying to survive winter, especially in the more northerly regions. With lives shortened by the mite, larger numbers of individual bees than usual die during the winter. This period then becomes even more stressful for the already weakened colony. It may die, or if it survives, it lacks vitality and may not be able to build up normally in the succeeding spring and summer.

Treatment. Two medications are available to protect colonies against tracheal mites — menthol and Miticur (amitraz). Both materials are effective when properly applied, although not totally so. Theoretically, it is possible to completely eliminate mites from a colony. Practically, such elimination is almost impossible, especially with heavy infestations.

Menthol, normally available in crystalline form, vaporizes in the hive. It kills adult mites, but it does not kill eggs or larvae. Therefore, menthol must be present for 14 continuous days (the development time of a mite) to kill all adult mites as they mature. It should be applied when temperatures are continuously warm, between 60°F (16°C) and 80°F (27°C), ideally in the 70°F (21°C) range. To avoid contamination of the honey crop, there can be no honey supers on the hive during the 14 days of treatment.

The standard dosage of menthol crystals for a 14-day treatment is 50 grams (1.8 ounces) per colony. When the temperature is below 80°F (27°C), the crystals are placed on top of the frames in the brood area, enclosed in an envelope of fine-mesh cloth or screening. At higher temperatures, the crystals should be placed on the bottom board. Menthol is very disruptive to the colony and even

POSSIBLE SYMPTOMS OF TRACHEAL MITES

Tracheal mites are microscopic. They cannot be seen with the naked eye. Most of their life cycle takes place within the host bee, so positive diagnosis of a tracheal infestation can only be done after dissection and microscopic inspection. However, there are some symptoms at the hive that suggest a tracheal mite infestation.

- Bees crawling about at the hive entrance, unable to fly.
- Poor clustering in cold weather. If the colony dies, bees may be found randomly throughout the hive bodies (instead of in a cluster), usually with plenty of honey on hand.
- K-wing, a phenomenon in which the afflicted bee's wings at rest are not folded along the abdomen but instead angle out to the sides, with the two wings on each side separated and forming the letter K with the bee's body.

when placed on the bottom board, in warmer temperatures (above 80°F or 27°C) the bees will be driven from the hive. Even at lesser temperatures, assume that when you treat, foraging and brood production will be reduced or stopped.

If you can smell the menthol, it is working. If the odor is faint or there is no odor, the menthol is not effective. Be careful. It can be harmful for people to inhale menthol. Whether in use or in storage, menthol will melt and run at 95° to 97°F (35° to 36°C).

Miticur is a contact miticide, available in the form of impregnated plastic strips, which are hung between the frames in the hive. The bees come in direct contact with the strips and Miticur rubs off on them. It in turn transfers to the mites, killing them.

The standard dose for Miticur is three strips per colony, applied after the honey supers have been removed, and maintained in the hive for at least 6 weeks. Miticur is not harmful to humans when applied according to the label instructions. However, it is a pesticide and care should be taken when handling it. Protective gloves are recommended.

Since neither treatment is likely to be 100% effective, annual treatment is necessary to prevent serious re-infestation. Although we can do a fair job of treating individual hives, treating a neighborhood or other geographic area ranges from difficult to impossible. Therefore, other hives in the area are a possible source of re-infestation.

Timing of treatment. Summer is the logical time to treat bees with menthol. For most of the country, it is when we can reasonably expect the requisite 14 days of continuously warm weather. However, it is the same period when honey supers will be on the hive. As with any medication, no treatment should occur while supers are in place. As an alternative treatment, late summer or early fall, just after the crop has been removed, has proven to be reasonably effective. However, because there are unlikely to be 14 continuously warm days at that time, not all the mites will be killed, although the mite population will be reduced and the colony should be able to survive the winter.

Miticur does not need warm temperatures to be effective. It may be applied after the supers come off at the end of the season.

To further protect our honey from possible contamination, neither medication should be applied during the 6 weeks just prior to installing supers. For both menthol and Miticur, read the labels carefully each time you acquire these or any other material to be applied in the hive. Keep in mind that formulations, recommendations, and labelling information may change over time.

Collection and diagnosis of bees. If there is a facility available nearby that will do mite inspection, it is important to take a large enough sample of live bees, and to take the bees from the proper place in the hive.

A sample 50 bees is the minimum, although more bees are better. Obtain the older bees from the honey supers, if any are present, or, from locations outside the brood nest. Collect from two or more locations in the hive. If the bees must be kept for a time, immerse them in alcohol. Rubbing alcohol is acceptable. If the colony has

CONTROLLING MITE TRANSMISSION

When tracheal mites were first discovered on this side of the Atlantic, there was some discussion of restricting and regulating bee movement. Now, any program to certify bees as being tracheal mite-free or to restrict the movement of bees interstate has largely been abandoned. It is accepted that the mite is everywhere and not controllable except by individual beekeepers. It is up to each beekeeper to take care of his or her own colonies on a continuing basis. The mites have not acted the same or responded to treatment in the same manner everywhere. This is presumably because of the two races or strains. Therefore, no one yet has all the answers. It is imperative that we not become impatient and take actions to control these mites that are untried or illegal. It is too easy to cause resistance to treatment and to contaminate honey.

died and only the dead bees are available for diagnosis, collect the freshest bees possible and immerse them in alcohol.

The Varroa Mite

The Varroa mite (*Varroa jacobsonii*) originated in Southeast Asia where it is a parasite of the Eastern honey bee, *Apis cerana*. It was first discovered on the Western honey bee, *Apis mellifera*, in 1960. The crossover resulted from beekeepers intermingling the two species and further spread has been encouraged by beekeepers transporting colonies.

This mite is now found on every continent except Australia. No one is sure how it came into the United States, but it is most likely that it arrived with queen bees that were brought in illegally. By 1992, Varroa mites were found in at least 40 of the United States and they have continued to spread. As with the tracheal mite, the spread has been furthered by movement of migratory hives and package bees.

Life cycle of the Varroa mite. The numbers of mites found in a colony of honey bees varies with the season. The fewest mites are found in spring, increasing over the summer to a high point in the fall, then falling off over winter to a low point in the spring. During the spring and summer, the mites are found mostly on brood. In fall and winter, they are mostly on adult bees.

Within this cycle, only adult female mites are found outside the cells of a beehive. The mite's life cycle begins when an adult female mite leaves a host bee and enters a brood cell as that host bee is feeding a young larva. Though the mite is found on all three castes of the honey bee, its preference is for drone brood; the mites enter drone cells in greater numbers than they enter worker cells. Once the mite is in the brood cell she hides in the brood food. As many as twenty-one mites have been found hiding in a single cell, though smaller numbers are more common. The mite is liberated from the brood food as the food is eaten by the bee larva. Soon after the cell is capped the mite moves to the prepupa and begins feeding. Sixty hours after the cell is sealed the mite lays her first egg, with succeeding eggs at approximately 30-hour intervals. New young mites reach maturity within the brood cell, where the mites then mate. The male mite then dies and the newly mated female leaves the cell with the emerging bee, moving to a new host bee to continue the cycle.

Mites feed on adult bees by puncturing the abdominal integument and sucking hemolymph (blood). Though not actually feeding, the mites move to secluded locations on the bee's body where they are difficult to detect visually. During the feeding stage mites may transfer to other adult bees as the bees brush against each other in the hive. The feeding stage may last for several days.

The total development time for an individual mite from newly laid egg through maturity and mating is 10 ½ days. Most of the races of *Apis mellifera* are good hosts for the Varroa mite because of the length of the bees' pupal or postcapping development time, which is typically about 12 days. In this time at least one mite can come to maturity in each cell. However, some races of *A. mellifera* (e.g., the African bee, *A.m. scutellata* and the Cape bee, *A.m. capensis*) have a shorter development time and fewer mites can mature.

Colony damage and mortality. Individual bees infested with Varroa are harmed in two ways: first, by loss of hemolymph, which in itself is serious, and second, by the puncture wound which allows entry of infections and disease. Even in small infestations, bees suffer weight loss and shortened life. If the per bee infestation is less than six mites, the bee usually reaches maturity. Developing mites, therefore, also reach maturity. However, the adult bees are weakened and their lives are shortened by as much as 50 percent. Bees infested with even a single mite during the brood stage can lose 6 percent to 7 percent of their adult weight. Bees infested with six to eight mites during the brood stage can lose 25 percent of their adult weight. However, many of these pupae never mature to emerge.

Other damage includes asymmetrical wings, misshapen legs, and shortened abdomens. Drones have a reduced number of spermatozoa, reduced weight, and lesser frequency of flight activity.

The sum effect on a colony infested with Varroa mites is serious harm and if no control measures are taken the colony will die. Without beekeeper intervention, the probability of mortality is 10 to 15 percent the first year, 20 to 30 percent the second year, and perhaps 100 percent in the third year. At the most, an untreated colony will live only five years after infestation.

Detecting Varroa Mites

Finding mites by inspecting individual live bees is usually not successful. The mites are very difficult to see on the bees, even with several mites present. They blend in with the bee, and when not actively feeding, they hide. Some method of removing the mites from the bees is necessary. Even then, mites may be difficult to find in a light infestation.

Inspecting adult bees. Following are two methods for testing adult bees for mites. These methods are most effective in the late season when mites are more concentrated on the adult bees, although they do not give good results if the infestation is light.

#1. The "ether roll" is relatively simple and uses a minimum of equipment. A glass jar of about 12- to 16-ounce capacity with a tightly fitting lid is required, and a pressure can of ether (such as

One way of detecting mites is to use a capping scratcher to remove and uncap drone brood. The older pupae are easy to remove, and once they are uncapped, mites on their bodies are easily visible.

automobile starter fluid). Place at least 500 live bees in the jar. Crack the lid and give them a one-second shot of ether. Seal, tilt the jar on its side, and roll the contents. Moisture will form on the inside of the jar and after 20 to 30 seconds, if mites are present, they will appear adhering to the film of moisture as the jar is rolled. Do this test quickly, then dump the bees on a clean exposed surface. They may recover.

#2. The shaking method is similar to the ether roll. Place a quantity of bees in a jar of alcohol, detergent, diesel fuel, gasoline, or even hot water. Shake vigorously for about 1 minute. If mites are present on the bees, they will fall off. Sieve out the bees and then strain the liquid to look for mites.

Inspecting brood. Mites are easier to detect on brood, especially on drone brood. Inspection is best done by uncapping and removing some of the drone pupae. Very young pupae are difficult to remove; they break up easily. Older pupae may be removed from their cells with a capping scratcher. These older pupae may be identified by their eyes, which have color. The younger pupae have little or no eye pigment.

Once the cells of the older pupae have been located, slide the points of the capping scratcher horizontally under the cell cappings so as to impale a number of brood at one time. Lift the cappings and the brood from the cells to inspect. If mites are present, they should be obvious on the white or light colored bodies of the pupae.

Examining a few cells is not enough. A casual inspection should include at least 200 cells (about 13 square inches), but at the height of colony development, about 450 cells (about 28 square inches) need to be examined to identify a mite infestation of 1 percent at an accuracy of 99 percent.

A way to help mite detection is to place a frame of drone foundation in the brood area. This will serve to concentrate a large number of drone brood in one place. Drone foundation is available from some bee supply sources. Otherwise, you can place some damaged comb in the brood area, which the bees will probably rebuild as drone comb.

Inspecting the hive. It is possible to inspect for mites without handling bees or brood directly. This is done by using bottom board inserts to collect mites that are killed or stunned by some suitable treatment applied to the entire hive. The inserts have two parts: a sheet of sticky paper or cardboard that covers the entire bottom board, and a screen that covers the sticky paper and serves to keep bees and larger hive debris off. The screen is raised slightly above the sticky paper and is sized to prevent the bees from passing through. Mites, which are killed or stunned by appropriate hive treatment, fall to the bottom of the hive and through the screen, where they are caught on the paper. Apistan (fluvalinate) and Miticur (amitraz) are the two approved treatment materials. Both come in the form of impregnated plastic strips that are placed in the hive for a prescribed period. Both are contact miticides and do not harm the bees when used properly. Directions for use are carried on the labels and they must be used only in accordance with these directions. Miticur was discussed earlier under tracheal mites. The same methods of treatment and cautions pertain when using this material against the Varroa mite. Of course, using Miticur against either mite automatically protects against the other.

Bottom board inserts can also be used without medicated strips. When so used they will serve to retain any mites that die naturally in the hive and fall from the bees. They are most effective toward the end of winter, before the start of the active season. The screen should be of fine enough mesh to pass mites while retaining the bulk of other debris, but at the same time the debris should not be allowed to accumulate and prevent the mites from falling through.

Tobacco smoke. Tobacco smoke may be used in conjunction with inserts and screens. About 2 to 3 grams of tobacco should be ignited in the smoker (without other fuel present) and blown into the hive in the evening after the bees have stopped flying. The entrances should then be blocked with newspaper or other material to contain the bees and the smoke. The hive may be opened in the morning and the bottom board insert examined.

Apistan and Miticur. Although the label directions may allow, it is not advisable to use either as a routine prevention treatment against Varroa. Reserve these materials for treatment of an *actual* Varroa infestation. This reduces the possibility of mites establishing a tolerance for the medication. After an infestation is detected, use either material strictly in accordance with the label directions.

Other treatment medications and materials have been discussed in beekeeping literature, as have other methods of treatment. It is important to remember that unauthorized, untested, or otherwise unproven treatment can be detrimental to bees and often to humans. Further, misuse of approved or experimental materials can lead to the development of resistance to treatment in the mites. With only one or two chemicals available for treatment, we can not afford to lose any treatment because of beekeeper carelessness or ineptitude. Always use only approved materials in accordance with the label instructions.

Nonproblem Creatures for Honey Bees

Let's look first at some other creatures that live in the hive, creatures that may be characterized as nonproblems — earwigs, spiders, and some ants. Earwigs and ants are often found living on the inner cover; spiders sometimes are found under the rim of the outer cover and elsewhere around the hive. Each of these is interested only in a dry, secure space to hang out. They do little harm to the bees or the honey. For the most part, they feed outside of the hive.

Yellow jackets, which are wasps, are often found in and about the hive in the late season and seem primarily interested in stealing honey. Rarely have I ever found more than one or two at a time, and their impact on the colony is negligible. However, in some parts of the world, certain species of yellow jackets and other wasps are a serious problem for bee colonies, stealing large amounts of brood as food for their own developing young, often killing bees and decimating the population in the process.

Opossums are sometimes found nosing around a hive and have been reported as pests, but apparently they do no real damage.

Occasionally, an individual bird, such as a kingbird, is reported to be preying on flying bees in the vicinity of a beeyard. In most

PESTS AND PREDATORS OF HONEY BEES

Many insects, birds, and animals have at one time or another been considered pests or predators of bees, but the actual number that are of any real consequence in North America is quite small. The list includes bears, skunks, mice, wax moths, and in some parts of the United States, certain ants. Other insects and animals may be found in and about the hive but are rarely considered a problem. In this category are earwigs, spiders, some ants, yellow jackets, opossums, and birds such as martins and kingbirds.

instances, little can be done, and except to a queen breeder, no significant damage is done to the colonies.

Problem Creatures for Honey Bees

Ants are mentioned earlier as sometimes living in a hive, but being no real problem. However, in some areas, notably the South, particular species of ants are a serious problem and must be dealt with. These ants invade bee colonies and can cause their demise. Raising the hive on legs and placing the legs in cans of oil is an often practiced defense. Alternatively, the legs can be painted with one of the available sticky substances used to trap insects. Both materials need to be watched and the sticky material especially needs to be renewed periodically. Other solutions to the ant problem may be used locally. If you have these ants, ask around.

    Wax moths can be a serious problem. However, they are rarely a problem to an otherwise healthy colony. The moths are successful only when a colony has problems that prevent it from policing the hive adequately. Although moths may enter and lay eggs in a normal healthy colony, the bees remove them or the resulting larvae. If the hive is weakened though, the moths are able to get ahead of the bees, laying eggs that successfully hatch. The resulting larvae burrow through the comb, eating impurities in the wax and leaving

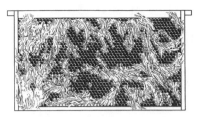

Wax moths are a particular threat to weak bee colonies or empty equipment, where they may burrow into the comb, eat impurities in the wax, and leave behind extensive webbing and frass.

behind large amounts of webbing and frass. They damage the comb, destroying it completely if unchecked, and sometimes cause the death of maturing brood which become trapped in their cells and are unable to emerge as adults. The bees are largely incapable of coping with the wax moth once it has successfully become established.

Occasionally, wax moths will be found in empty supers that have been placed on a hive optimistically. If the supers are not really needed, there is no nectar flow, for instance, or the colony is too small and does not need that much storage space, wax moths may move in to the unoccupied space. In this situation, a problem exists. They can take over and become difficult or impossible for the bees to remove.

The defense against wax moths in the hive is to maintain the colony in vigorous good health through good hive management. Wax moths also invade comb in empty equipment off the hive. In fact, more damage is probably done to empty equipment in storage or to abandoned equipment than to hives occupied by bees. Care should be taken to protect such equipment during the off season. For this purpose, paradichlorobenzine (PDB), a form of moth crystals, is available. Use only PDB for such protection. Another form of moth protection commonly available, napthalene, should not be used. Residual napthalene in the comb can harm bees. PDB has no effect on bees if the comb is aired for 24 hours or so before use.

To protect unused equipment containing comb, stack the supers or hive bodies on a solid, even surface that will prevent rodents from entering at the bottom. Counting up from the bottom of the stack, place a sheet of paper about 6 to 8 inches square on the top bars of the frames of every third or fourth super. Deposit about ⅓ cup of PDB crystals on each piece of paper. Cover the top of the stack. A spare outer cover is ideal, but anything to keep out rodents will do. Over time the PDB crystals will vaporize. The vapors, being heavier than air, will settle through the stack. For long-term storage, the PDB may need to be renewed periodically.

Because wax moths are killed by freezing temperatures, they are less of a problem in the more northern parts of the country. But protect against them wherever you live.

Mice find a beehive to be a wonderful place to spend the winter and they can cause a great deal of damage. You should assume that a mouse will find your hive at some time or another, so be prepared. Keep them out.

The mouse enters on cool nights in the fall when the bees are clustered. It chews a hole several inches in diameter and builds a spherical nest through several frames of comb. The nest is made of materials the mouse brings in—grass, leaves, and other typical mouse nest material. The mouse apparently builds this nest at a time when it is too cold for the bees to break their cluster. Once complete, the mouse can stop up its entrance to keep bees out if the weather warms up.

In the worst case situation, a mouse can create enough disruption to cause the colony to abscond. Occasionally, the bees are able to attack the mouse and kill it. Since they cannot move it, they then may embalm it with propolis.

The best defense against mice is a metal guard over the entrance. Mice can chew through a wooden entrance reducer or most other wooden barriers. A strip cut from ⅜-inch wire mesh will serve, or obtain one of the metal guards made for the purpose.

Don't delay in placing that mouse guard in the fall. Late summer may even be better—mice get ready for winter early.

Skunks enjoy a meal of adult bees, which they catch alive at the hive entrance. They are seemingly impervious to bee stings and at times are a serious pest of beehives. Over a period of days, a skunk can severely weaken a colony.

The skunk approaches the hive entrance at night, scratches on the landing board, and catches individual bees as they come out to investigate. Each bee is caught in the skunk's paw, rolled on the ground and killed, and then popped into the skunk's mouth.

There are several kinds of evidence of a skunk's depredations. One sign is the holes it digs around the yard as it feeds on grubs in your lawn. Then, just in front of the hive entrance you may see that the ground is bare where the skunk rolled the bees around. The landing board may have dirt and mud on it from the skunk's paw as it reached for the bees. On the ground in the general area, you may find little piles of dead bee remains. After the skunk has chewed the goodness out of a wad of bees, it may regurgitate them as a blackish mass.

Another possible signal of skunk depredations is a sudden display of testiness in your bees. Constant harassment quickly lowers their threshold of irritability. If you have any reason to believe you have a skunk problem, take immediate steps.

The most common defense against skunks is a guard placed on the ground in front of the hive entrance that will give the skunk uncertain or uncomfortable footing. A piece of rolled poultry netting or other fencing material, or a thin shingle studded with tacks protruding upwards, are two possibilities.

Bears are of little consequence to most beekeepers. However, for that small number of beekeepers who live in bear country, they can be a major problem. The contents of a beehive are choice food for a bear and they enjoy both the honey and the brood. Probably a bear sees a hive as just another tree stump full of grubs, with some sweetening to boot.

Once a bear has discovered a hive or a beeyard, it is hard to discourage its attraction. The bear is likely to return on successive nights until little is left intact. The hive is usually destroyed, with much of the woodenware broken up or even carried away.

It can be difficult to defend against bears. They are strong and persistent and often ignore people, barking dogs, and other disturbances. They are able to pull apart hives even when the equipment has been stapled or strapped together and will climb over or around most obstacles to get to them. Beekeepers have resorted to putting hives on platforms, inside of abandoned vehicles, in buildings, and behind extensive fences or barriers. It is possible to make a hive or a yard bear-proof, but it can become prohibitively expensive and often also make the hive beekeeper-proof. Imagine climbing a ladder every time you want to work your hive, carrying hive bodies or full supers up and down, or consider the problems in trying to open a hive inside an old car.

The recommended defense is to erect an electric fence, before the bear ever finds the hive. Normally, a bear will be deterred by an electric fence if it is just in an investigative mood. However, if the bear has already been into a hive, and then the fence goes up afterward, the bear may go through it on the return visit. Some bears apparently consider the reward to be worth the shock. If there is any suspicion that bears are in your area, think seriously about erecting an electric fence.

The Africanized Bee

In 1957, a number of queen bees were taken from Africa to Brazil to be used in an experimental breeding program. The goal was to breed some of the supposedly desirable characteristics of these African bees into the stock then existing in Brazil. These African bees were the same species, *Apis mellifera*, that were and are now commonly kept by beekeepers in both North America and South America. They were, however, a different subspecies than any then being kept in the Western Hemisphere. These bees were *Apis mellifera scutellata*, African bees, known for their low threshold of tolerance for other living creatures. The African bee evolved in a particular climate and environment in southern Africa under conditions that encouraged the development of a number of highly undesirable characteristics and traits, excessive stinging and excessive swarming being two of these.

During the experimental period in Brazil, these African bees were kept under careful control and not allowed to swarm. The researchers involved recognized the problems that could arise if this subspecies was released into the wild, whether deliberately or by chance. Unfortunately, through an error, the controls were relaxed and these bees did swarm. More than twenty of these swarms with pure African queens escaped into the wild.

In the more than 35 years since, the offspring of these original swarms have been spreading and interbreeding, moving through South America into Central America, and in 1991, into the United States. Of course, beekeepers have been keeping bees in South and Central America for about as long as we have been keeping them in North America. They, too, keep the various European races of *Apis*

WHAT CAN YOU DO ABOUT AFRICANIZED BEES?

First, recognize that we will not truly know how to deal with these bees until they arrive and see how they acclimatize. Furthermore, we don't truly know when they will arrive in any given area. They are credited with moving naturally about 300 miles per year. Beekeeper involvement could inadvertently hasten the spread as colonies are moved about the country for pollination, and media hype will hasten the perception of their presence. Individual states are working on plans to control and to deal with these bees, and although some coordination between states has begun, much more needs to be done. Many beekeepers and appropriate authorities have not yet accepted the potential scope of the problem, and feet are dragging.

On a personal level, start by reading. Learn as much as you can about these bees, their history, and the work that has been done with them since they came to the Western Hemisphere. Stay current. Many beekeepers, researchers, educators, and public officials are working on the problem. New information is being discovered and published regularly.

Help with the education process. Other beekeepers, the public, and officials at all levels of government need to be informed. Because of their pollination activities, honey bees are too important to our way of life to allow unnecessary and unreasonable restrictions to be put in place. If your state or municipality does not have a plan for coping with the Africanized bee, help develop and put one in place. Don't wait for others to do it. It may happen too late, and it may be the wrong plan. If beekeepers don't stay in the forefront on this, we may have inappropriate plans imposed on us.

(Continued on page 165)

If you live in an area where the Africanized bee has already arrived, find out what plans, restrictions, and regulations are in place. Abide by them. They are for the benefit of everyone. If you feel that unreasonable restrictions have been placed, work toward changing them.

In dealing with bees in an area where the Africanized bee has arrived or is suspected, common sense applies. Swarms from unknown sources must be suspect and should be destroyed. Keep your stock pure by requeening regularly with a marked queen of known breeding and quality. Know the other beekeepers in your area and know their practices. Be prepared to compensate, if they do not take their responsibilities seriously.

The problems associated with the Africanized bee are serious. They cannot be ignored.

mellifera — Italians, Caucasiana, Carniolans, and the like. The resulting crossbreeds of these with the African bee have been commonly termed the Africanized bee. (Crossbreeding takes place when virgin queens fly out to mate and encounter Africanized drones.)

One of the most difficult aspects of dealing with this crossbred bee is that the negative characteristics of the African bee are dominant. Compared to most other races of bees, both the African and the Africanized bee react to disturbances much more quickly, sting more readily and more persistently, and swarm more often. Individual bees differ very little from European bees. They are no bigger and are, perhaps, even smaller. Each bee can sting only once, and its venom is no more potent than that of a European bee. The difference is in intensity: Africanized bees have a lower threshold of reaction, larger numbers sting in a given instance, and they follow their victims greater distances from the hive.

Over the 30-some years that these bees have been wending their way north, stories have abounded. We have heard that they are mean and vicious, poor pollinators, poor honey producers, and simply impossible to be around. We have also heard that they are gentle, easy to work with, good honey producers, and adequate pollinators. These reports have come from an interesting cross section of beekeepers, researchers, public officials, and the public. Over the years, a number of people and small animals have died as a result of stinging incidents involving these bees. At the same time, there are instances of people living and working in close proximity to Africanized colonies.

A part of the problem is that Africanized bees act differently under different circumstances. As with any race of *Apis mellifera,* a small or newly established colony of Africanized bees is relatively gentle and easy to deal with. Later, as the colony matures and gains confidence, it may become more aggressive. At the leading edge of the movement of the Africanized bees, where they are contacting and interbreeding with European bees, and where the Europeans may still be a strong element in the mix, behavior is probably differ-ent than it will be later on — when the Africanized influence be-comes more dominant.

And finally, in areas where the Africanized bee has become established, both beekeepers and the public are learning more about living and dealing with it. Incidents do occur, but they have become somewhat more routine to cope with, although they will probably never be completely routine.

What does all of this mean to us in the United States and Canada? We don't really know. Physical barriers to stop the movement north of these bees have been attempted and failed. They have entered Texas and continue to spread. African bees evolved in a warm climate. There is strong evidence and supposition that they will not survive the winters of the northern United States and will be con-fined to a tier of southern states. At the same time, one school of thought believes they can survive as far north as Minnesota.

Even if Africanized bees do not move north permanently, there is some possibility that they will move through inadvertent bee-keeper movement during the spring and summer months, and cause problems until they are wiped out by the cold of winter.

Such incidents, or media reports of incidents in the southern United States, could panic the public and, in turn, public officials. A very possible outcome may be the imposition of regulations and restrictions on the beekeeping community that reflect this panic, rather than the realities of the situation.

Pesticides

Rarely does anyone deliberately spray insecticide or pesticide on honey bees. Most people recognize their value. However, opportunities abound for bees to be exposed to pesticides — orchards, farms, and home gardens, to name a few. The number of available insecticides, herbicides, fungicides, and miticides is large, and their effects on honey bees are quite varied. Some are benign, some are lethal, and some fall in the middle, with their effects depending on formulation, time of day of application, or time elapsed since the application. Of most concern, obviously, are the insecticides.

If insecticides that are harmful to bees are being used, it can be difficult to defend against them, primarily for lack of knowledge of their source. Accepting a normal foraging range of up to 2 miles for a colony of bees, this is an area of over 12 square miles if the bees exploit the range to its fullest. A foraging distance of 3 miles is well within their capability, increasing the area to about 28 square miles. A large amount of insecticide may be applied within such an area without the beekeeper being aware of it.

A primary defense against spray damage is avoidance. If there is a choice, do not keep bees in an area where insecticides are used regularly. For instance, orchards are very tempting locations to establish a beeyard, but orchards usually are recipients of a variety of sprays. Any colonies located there cannot help but suffer. Corn is another crop to be avoided. Corn pollen is very attractive to bees: they quickly find any growing in their foraging range. Sweet corn is usually sprayed with insecticides at about the same time it is yielding pollen.

At times, contact with insecticides is unavoidable — a one-time spraying against a specific pest, for instance. Given notice of an imminent spray in the neighborhood, there are steps a beekeeper can take. One obvious step is to temporarily move the hive to another neighborhood. However, this can be inconvenient, and sometimes impossible, on short notice. An alternative is to confine the bees so they can not fly out to be exposed.

Most spraying takes place during the growing season, when the weather is warm. Simply closing up the hive so the bees cannot leave could suffocate them. Instead, cover the hive with a large sheet of burlap or similar open-weave material. Drape it loosely but be sure it is in close contact with the ground. Allow the bees to leave the hive but do not allow them to get out from under the burlap. Dampen the burlap periodically. The evaporation of moisture helps cool the hive. A colony can be kept confined in this manner for 2 or 3 days, given food reserves in the hive and if moisture is available.

Even with the best of care, a hive may encounter insecticides. Depending on the specific material involved, the effects on the colony may vary. Some materials are lethal to bees on contact. The foragers die in the field and do not harm the hive directly. Of course, the population may be devastated. Other materials may contaminate nectar or pollen at the source and not harm the foragers directly, but may kill bees in the hive as they eat the contaminated material that is carried back.

A hive suffering from insecticide damage reflects the particular mode of contamination. A sudden drop in field bee population suggests that bees are being killed in the field. Bees dead at the hive have probably eaten contaminated stores. In this case, the number of bees dying often overwhelms the ability of the hive to cope. Large numbers of dead bees may pile up at the hive entrance, whereas stricken but still-live bees act abnormally by crawling aimlessly, unable to fly, and sometimes showing paralysis.

Colonies contaminated by insecticide have an uncertain future. Some that are lightly stricken will survive and be back to normal fairly soon. Heavily contaminated colonies die quickly. In between, some may hang on for a while but ultimately die, either as a direct or indirect result of the contamination.

If a colony has been stricken with insecticide poisoning but remains alive, try to discover the source and the particular insecticide involved. If there is any reason to suspect that contaminated stores remain in the hive, a prudent step is to remove and to destroy the contaminated comb. Feeding in such circumstances may also be in order. Presumably, population has been reduced and consequently, foraging capability is reduced as well.

Publications About Beekeeping

Books about bees and beekeeping abound, and new ones appear regularly. As a beginner, you neither can nor would want to read them all. The list below will help you to pick and choose.

This list emphasizes recent books. Many advances in the knowledge of bees and beekeeping have been made over the years. Especially in recent years, we have seen a great increase in the knowledge of bee biology, which has allowed us a better understanding of their activities and behavior. This understanding, in turn, has helped us to keep and to manage bees more effectively. The older books, even those written as recently as 20 years ago, may not reflect some of this important new knowledge.

In selecting books, you should realize that beekeeping, in general, is the same worldwide. However, there are national and regional differences in both equipment and methodology. Beginners are best served by books written in their own country.

Novices are often at a loss as to where to locate beekeeping books. The public library is an obvious source. The holdings may be limited, although many libraries have a surprisingly good selection of beekeeping books. Most bookstores stock only a limited variety of beekeeping titles, but almost any bookstore can order a book for you if given the title and the names of the author and the publisher. Several sources that offer beekeeping books are shown below.

Following is a list of selected books to supplement the material in this volume. Unfortunately, several of these books are now, or may soon be, out of print and not readily available through normal channels. Try the book sources listed. This list is not intended to be all inclusive. It is but a starting place. There are many other good books.

General Reference

✖ Dadant and Sons, ed., *The Hive and the Honey Bee*, rev. ed., Dadant and Sons, Hamilton, Illinois, 1992.

✖ Morse, Roger A., and Ted Hooper, eds., *The Illustrated Encyclopedia of Beekeeping*, E.P. Dutton Inc., New York, 1985.

✖ Root, A. I., *The ABC and XYZ of Bee Culture*, 40th ed., A.I. Root Co., Medina, Ohio, 1990.

Practical Beekeeping

✖ Bonney, Richard E., *Hive Management*, Garden Way Publishing, Pownal, Vermont, 1990.

✖ Morse, Roger A., *The Complete Guide to Beekeeping*, 3rd ed., E.P. Dutton, New York, 1986.

✖ Sammataro, Diana, and Alphonse Avitabile, *The Beekeeper's Handbook*, 2nd ed., Macmillan, New York, 1986.

✖ Taylor, Richard, *The How-To-Do-It Book of Beekeeping*, 3rd ed., Linden Books, Interlaken, New York, 1980.

Biology, Organization, and Communications

✖ Free, John B., *The Social Organization of Honeybees*, Edward Arnold Ltd., London, 1977.

✖ Frisch, Karl von, *Bees: Their Vision, Chemical Sense, and Language*, rev. ed., Cornell University Press, Ithaca, New York, 1971.

✖ Seeley, Thomas D., *Honeybee Ecology*, Princeton University Press, Princeton, New Jersey, 1985.

✖ Winston, Mark, *The Biology of the Honey Bee*, Harvard University Press, Cambridge, Massachusetts, 1987.

Diseases

✖ Bailey, Leslie, and B. Ball, *Honey Bee Pathology*, 2nd edition, Academic Press, New York, 1991.

✖ Hansen, Henrik, *Honey Bee Brood Diseases*, Wicwas Press, Cheshire, Connecticut.

✖ Morse, Roger A., ed., *Honey Bee Pests, Predators, and Diseases*, Cornell University Press, Ithaca, New York, 1978.

Historical Interest

✖ Langstroth, L.L., *Langstroth on the Hive and the Honey Bee*, A. I. Root Co., Medina, Ohio, 1853, reprinted 1977.

✖ Miller, C. C., *Fifty Years Among the Bees*, Molly Yes Press, New Berlin, New York, 1915, reprinted 1980.

Book Sources

Catalogs are available from all of these sources.

✖ Dadant and Sons, 51 South 2nd Street, Hamilton, Illinois 62341
✖ International Bee Research Association, 18 North Rd., Cardiff CF1 3DY, United Kingdom
✖ A. I. Root Co., 623 West Liberty Street, Medina, Ohio 44256
✖ Storey Communications, Inc., 105 Schoolhouse Road, Pownal, Vermont 05261
✖ Wicwas Press, P.O. Box 817, Cheshire, Connecticut 06410-0817

Publications

Two monthly national magazines are available in the United States, both of which can be read profitably by beginning beekeepers. Write to them for subscription information.

✖ *American Bee Journal*, 51 South 2nd Street, Hamilton, Illinois 62341
✖ *Bee Culture*, 623 West Liberty Street, Medina, Ohio 44256

SOURCES

Where to buy equipment and bees is often a mystery for beginners. One or more dealers may be within reasonable distance of where you live, but whether local, regional, or national, they rarely advertise outside of beekeeping circles. Query local beekeepers as a starting place. Beyond this, there are several mailorder sources whose advertisements run regularly in beekeeping publications. Several of these are listed here. Write for their catalogs.

✖ Betterbee, Inc., 8 Meadow Road, Greenwich, New York 12834
   (800) 632-3379

✖ Brushy Mountain Bee Farm, 610 Bethany Church Road, Moravian Falls,
   North Carolina 28654 (800) 233-7929

✖ Dadant and Sons, Inc., 51 South 2nd Street, Hamilton, Illinois 62341
   (800) 637-7468

✖ Jones, F. W. & Son Ltd., 68 Tycos Drive, Toronto, Ontario, Canada JOJ 1A0

✖ Kelley, Walter T. Co., 3107 Elizabethtown Road, Clarkson, Kentucky 42726

✖ Mann Lake Supply, 501 South 1st Street, Hackensack, Minnesota 56452
   (800) 233-6663

✖ Rossman Apiaries, P.O. Box 909, Moultrie, Georgia 31776 (800) 333-7677

You will also want bees, of course. If your choice is package bees, they can sometimes be obtained through a local equipment dealer or you may deal directly with the source and have them mailed to you. There are at least several dozen queen and package producers in this country. Ask for recommendations from other beekeepers. Look for the producers' advertisements in beekeeping publications.

Nucleus hives are usually available most readily from local sources. Although nucs can be transported, they do not travel as well as packages and they are not something to be shipped through the mail. Check with local dealers and with other bee keepers.

GLOSSARY

Abdomen The third section of a bee's body, containing such organs as the honey stomach, the digestive system, the reproductive system, and the stinger.

Absconding The act in which an entire colony leaves its hive to take up a new residence, usually caused by lack of food, excessive harassment, or other undesirable conditions. See also *swarming*.

Acarapis woodii Scientific name of the tracheal mite that infests the honey bee.

African honey bee Common name of *Apis mellifera scutellata,* the subspecies of honey bee brought to Brazil in 1957 for research purposes.

American foulbrood (AFB) A disease of bacterial origin that can infect the brood stage of honey bees.

Apis mellifera Scientific name for the European or Western honey bee, the bee commonly kept by beekeepers.

Apiary A beeyard; a place where one or more hives of bees are kept.

Bee bread Pollen as collected by the bee, mixed with small amounts of honey and stored in the hive.

Bee space The open space that bees maintain throughout the hive to provide passageways through the combs. Bee space is about ¼ - to ⅜-inches wide.

Brood Collectively, the immature stages of the bee, including eggs larvae, and pupae.

Brood chamber A section of the hive used by the bees for brood rearing and storage of their food reserves.

Brood nest The specific part of the brood chamber in which brood is raised.

Capped brood The portion of the brood that has been covered with wax cappings; specifically, the pupae.

Chalkbrood A fungal infection that can infect honey bee brood.

Comb The completed cells in a frame or in the hive where the bees raise brood and store food.

Emergence The birth of an adult bee; the point in a bee's life cycle where it completes its pupal development and leaves its natal cell as an adult.

European foulbrood (EFB) A disease of bacterial origin that can infect the brood stage of honey bees.

Foundation The thin sheets of pure beeswax embossed with hexagons given to the bees in frames as a base for their comb.

Honey The sweet material produced by bees from the nectar of flowers and stored in the hive as their food reserve.

Honey stomach An organ in the bee's abdomen in which is carried honey, nectar, water, or any liquid that the bee is transporting. Material in the honey stomach may be regurgitated or passed on into the bee's digestive system.

Hypopharyngeal glands Glands in a worker bee's head that secretes royal jelly or bee milk.

Larva (larvae) The second or feeding stage of the brood development cycle.

Mite See *Acarapis woodi* and *Varroa jacobsonii*.

Nectar Sugary solution secreted by flowers and collected by bees to be converted into honey.

Nucleus hive (nuc) A small but complete colony, usually of three to five frames, containing brood, adult bees, food stores, and a queen. Often used as a starter unit.

Nosema A protozoan infection of the mid-gut of adult honey bees.

Open brood Brood that has not yet been capped with wax, specifically the eggs and larvae.

Package, package bees Screened container containing a specific amount of adult bees and a queen, sold and used for starting new colonies.

Parthenogenesis (virgin birth) Phenomenon by which certain male insects, including honey bees, develop from unfertilized eggs.

Pheromone A chemical substance secreted by bees and other animals that, when sensed by others of the same species, causes specific actions or responses.

Pollen Dustlike grains produced in the anthers of flowers as the male element of their reproductive system. These grains are collected and stored by bees as the protein source in their diet.

Propolis A resinous substance collected by bees from the bark and buds of certain trees, used for filling in cracks and crevices in the hive and for reinforcing comb.

Pupa (pupae) The third stage of the brood development cycle, after which the bee emerges as an adult.

Queen excluder A device used in the hive to restrict the passage of the queen. It also restricts the drones.

Sacbrood A viral infection of honey bee brood.

Spermatheca An organ within the queen which carries the millions of sperm received from drones on her mating flights.

Spiracle The openings on the sides of the thorax and abdomen through which a bee breathes.

Swarming The act in which a portion of a colony (a swarm), including the queen, leaves its home to establish a new home elsewhere. The parent colony, now reduced in population, raises a new queen and continues as before.

Thorax The second, or middle, division of an insect's body, coming between the head and the abdomen. Here are attached the legs and wings.

Tracheal tubes Breathing tubes, which open from the spiracles and divide and spread throughout the bee's body by a system of tracheoles.

Trophallaxis The exchange of food that constantly takes place among the bees of a colony. Trophallaxis is a means by which pheromonal and other information is spread within the hive.

Varroa jacobsonii Scientific name for the Varroa mite, an external parasite that infests honey bees.

INDEX

Nosema disease, 109, 147–48
Nucleus hives (nucs)
    acquiring, 75–76
    installing, 89–90
Nutrition, 49–51

O
Opossums, 158, 159

P
Package bees
    acquiring, 74–75
    installing, 83–89
Pails, plastic, 131
Paradichlorobenzine (PDB), 160–61
Parthenogenesis, 31
Pesticides, 167–69
Pheromones, 53
Pollen
    consumption of, 32, 49, 50–51
    feeding, 52, 117–18
    human use of, 138–39
    recognizing, 104
Population cycle, 42
Population growth, 91, 101–2
    importance of, in winter, 108–9
Porter bee escape, 124–25
Predators, 159–63
Propolis, 15, 32
    human use of, 139

Q
Queen(s)
    brood development, 30–31
    cage, 86-88
    cage candy, 87
    development of, 33
    excluder, 16, 72–73

installing, 87-88, 92
life span of, 29
physical characteristics of, 29
requeening, 109–10
role of, 28–29, 33

R
Radial extractors, 133
Regulations, 11–12
Requeening, 109–10
Round dance, 55
Royal jelly, 32, 49
    human use of, 139–40

S
Sacbrood, 146
Sideline operators, 17
Skeps, 13
Skunks, 161–62
Smell, sense of, 7, 36, 80
Smoker, 70
Spiders, 158
Spring management, 118–21
Starline, 26, 27
Stingers
    parts of, 10
    removing, 10
Stings
    allergic reactions, 7 8
    as a defensive behavior, 6, 7
    effects of, on bees, 9–10
    reactions to, 8
    remedies for, 10
    tolerance to, 8–9
Storage of honey, 134–35
Strainers, 130–31
Sugar, feeding, 51–52, 116, 117
Supering. *See* Honey supers

Swarm, acquiring bees by, 77
Swarming, 44
    controlling, 48
    defining, 45
    process, 46-48
    timing of, 46

T

Tangential extractors, 133
Terramycin, 143–44, 146
Tobacco smoke, use of, 157
Tracheal mites, 109, 148–53

U

Uncapping knives, 130

V

Vandalism, 60
Varroa jacobsonii, 148, 153
Varroa mites, 109, 148, 153–58
Veils, 71–72
Venom, bee, 140
Virgin birth, 31
Vitamins, 51

W

Wag-tail dance, 56–57
Wasps, 23, 158
Wax moths, 159–61
Wax secretion, 51
Western honey bee, 24, 153, 154
Windbreaks, 111–12
Winter
    activities in, 107, 114
    checking during, 114–16
    colony health in, 109
    feeding during, 116–18

food stores, 110–11
hive protection, 111–14
importance of population
    growth in, 108–9
requeening, 109–10
Workers
    brood development, 30–31
    development of, 31–2
    as field bees, 32, 40–41
    as house bees, 32, 38–40
    life span of, 30, 40–41
    physical characteristics of, 30
    role of, 30, 32, 37–41
    tasks of, 39

Y

Yellow jackets, 24, 158